THE RECLAIMERS

THE RECLAIMERS

A COMPLETE GUIDE TO SALVAGE

SALLY BEVAN

SPECIAL PHOTOGRAPHY
BY ANDREW MONTGOMERY

Foreword by Thornton Kay

HODDER &
STOUGHTON

CONTENTS

FOREWORD

In the past ten years, despite our efforts to be less wasteful, the amount we throw away has grown by fifty per cent. Every year, for example, we make billions of new bricks and throw away billions of reusable old ones. Twelve bricks embody the energy of a gallon of petrol, so saving old bricks helps reduce global warming.

Happily, creative reuse of architectural and garden salvage – long the domain of idiosyncratic stylists – is now being encouraged by governments, to help reduce the ever-increasing amounts of material being sent to landfill.

The recent trade in salvage came about because some people saw the potential of the things that were being thrown away. Andy Thornton in Yorkshire, Adrian Amos in London and Rick Knapp in Bath were pioneers in the 1970s, heroically rescuing and carving out a market for unloved fireplaces, joinery, garden statuary, stone and floorboards.

As our built heritage became more protected, less was jettisoned, so the predominant salvage type moved from Georgian to Victorian. Now 1950s retro is all the rage. For dealers, the name of the game is to find something no-one wants, save a load of it, create an appreciation for it and then try to make a market in it.

Reuse should begin at home, by saving everything possible. When your electrician or plumber removes floorboards and skirting, it should be done like an archaeological excavation. All boards should be photographed, numbered and carefully stored; all then carefully replaced afterwards. Floorboards should be washed and waxed, not sanded. This will conserve the patina of age and, more importantly, energy and resources.

So, to sum up, save what you can and reuse where you can, and be uncompromising in the face of recalcitrant builders, designers and planners. Appreciate the old materials and the skills that went into them, and give them the chance of a second life.

Thornton Kay
Senior partner, Salvo
www.salvo.co.uk

WELCOME TO SALVAGE!

If you've been glued to the television series *The Reclaimers*, chances are it will have fired a new passion for architectural antiques. This book is designed to complement the series, helping you to become a successful salvage hunter yourself – knowing how to spot a bargain, what to expect from salvage dealers, and how to use salvage in a fresh and exciting way. If you haven't seen the series, don't worry – this book is also designed for you and covers absolutely everything you need to know to get started.

If you've never been to a salvage yard, you're in for a real treat. Somewhere between a 'Hollywood prop department and an elephant burial ground', as salvage expert Thomas J. O'Gorman puts it, salvage yards are fascinating and endlessly surprising treasure-troves. Often rambling and chaotic, sometimes neat and museum-like, they're always packed with architectural gems from the past. Stone angels watch over carved marble fireplaces, red telephone boxes, Victorian street lamps and roll-top baths. From radiators to railings and gates to gazebos – these objects are rescued from buildings being demolished or refurbished in the hope that they can be given a new lease of life.

Whether you're restoring a house to its former glory or simply shopping for something a bit different, salvage offers endless opportunities for creative people who are keen to find their own individual style. No two salvaged interiors will ever look the same. Each piece of salvage has its own unique story to tell – whether it's a weathered gargoyle from a stately home or a 1950s motel sign – and, as often as not, reflects a time when craftsmanship, design and materials were of a much higher quality than today. Where else will you find solid mahogany worktops, oak doors or a solid brass shower at rock-bottom prices?

So how does this book work? The first part, the Introduction, is designed to give you a general overview of the world of salvage. Part two, Salvage, deals with the different types of salvage in more detail, and is ordered alphabetically for easy reference. Under each heading you'll find a brief history of the item to help you know what to look for, as well as where to source your salvage, and how to clean and restore it. You'll also find suggestions and inspiration, including Expert Top Tips, so you can get the most from salvage in your home and garden.

The third part, the Directory, is probably the most important. This contains listings for a large number of salvage yards, restorers, antiques dealers and other related professionals up and down the UK. Visit a few salvage yards in your area. Get to know what's available and what you should expect to pay. Do some background research. In no time at all, you'll be snapping up bargains, doing deals, and creating an extraordinary, adventurous and totally unique living space for you and your family.

INTRODUCTION

A BRIEF HISTORY OF SALVAGE

There's nothing new about salvage. For thousands of years enterprising builders and keen-eyed collectors have picked among the rubble for reusable building materials and the odd architectural gem.

Sometimes this was a matter of convenience and thrift. Quarrying for stone, for example, is expensive and hugely labour-intensive. If you have a derelict stone structure on your doorstep it makes good economic sense to reuse the materials. Iron Age roundhouses, for example, often only leave their foundations behind; any stones used in the construction were almost always purloined by the next group of settlers and built into a new structure. Archaeologists can have a tough time trying to date such hotchpotch buildings (but then again, an archaeologist would tell you that's half the fun).

In other times and places, architectural reclamation was a deliberate aesthetic choice, a respectful nod to the past. Roman emperors borrowed decorative bits from earlier structures to embellish their own extravagant building programmes, and in the sixth century the architect and city planner Anthemius incorporated over a hundred columns salvaged from ancient ruins to build St Sophia Cathedral in Istanbul, Turkey.

Medieval builders were also great scavengers, rebuilding and adapting earlier structures. Castles and churches often reused Roman remains procured from a nearby ruin. Much of St Albans Abbey was created in Norman times by builders who just helped themselves to hand-made bricks from the ruins of nearby Roman Verulamium.

Sometimes the use of reclaimed materials even verged on the flamboyant. St Andrew's Church in Wroxeter lies next to the ruins of the Roman city of

Viriconium. This extraordinary church, which gets a mention in the Domesday Book, boasts church gateposts made from reclaimed Roman columns and a font carved from an upside-down column base. The ancient stonemasons also pilfered Roman carvings of lions and set them into the wall in what has to be one of the earliest examples of retro chic. Centuries later, as an extra flourish, the church tower was adorned with decorations acquired from nearby Haughmond Abbey after its dissolution in 1539.

During the Renaissance, the use of architectural antiquities became a strong political and artistic statement. Italian scholars drew inspiration from the ancient civilizations of Greece and Rome, which, for them, represented a great time of reason, learning and high culture. The Italians were sitting on a wealth of classical texts and architecture – the broken and ivy-covered ruins of the Roman Empire. Buildings, sculptures and other artefacts that had survived over the centuries now became desirable souvenirs for anyone with half an interest in the classical world. Even artists and architects such as Donatello and Brunelleschi are reputed to have searched through the debris, cherry-picking pieces to take back home.

By the eighteenth and nineteenth centuries every young aristocrat worth his salt was arranging a 'Grand Tour' – a marathon cultural trek of ancient cities such as Athens, Naples and Rome with the express purpose of bringing back inspiration and artefacts. Statues, columns, marble, stone and wall friezes were shipped back to Britain and incorporated into the design schemes of large houses and gardens. As you can imagine, not everyone was delighted with this trend. Soon after the seventh Earl of Elgin hacked his Marbles from the Parthenon in 1801, for example, the Greeks began a persistent but (as yet) unsuccessful campaign to get them back.

By the middle of the twentieth century a new market for architectural antiques was emerging. After the Second World War vast amounts of objects were becoming available as grand house after grand house was closed and cleared. During the 1960s a few enlightened architects and house restorers had begun to ask for traditional building materials and domestic architectural antiques. Never ones to miss a trick, antique dealers soon set up large warehouses stocked with all manner of architectural salvage – baths, tiles, floorboards, fireplaces, staircases, panelling – much of which came from these stately homes as well as Victorian and Georgian houses that had been gutted of all original features during the 1950s or demolished as part of inner-city regeneration programmes. By the 1980s the modern salvage trade was thriving.

SALVAGE TODAY

The architectural salvage business has to be one of the commercial success stories of the late twentieth century. Currently worth a staggering £1 billion a year, the trade in reclaimed architecture is now truly international, with antiques and money flying freely between the continents.

As the rest of the antiques trade struggles to survive, with customers disappearing, shops closing and proprietors going bust, the architectural salvage business is ruddy with health. What makes the salvage industry even more of an inspiration as a business model is that it represents the ultimate fairy-tale success story of cash from trash.

It really was a case of one man's rubbish being another man's treasure. During the late 1960s and early 1970s, while demolition crews and over-enthusiastic DIYers were still throwing Victorian fireplaces, doors and baths on to skips, a few canny people spotted the potential market. Evan Blum, owner of Demolition Depot in New York City, remembers: 'I'd just go to the edge of a demolition site and haul off a doorway.' Today his business is booming, with such celebrity clients as Robert De Niro and Isabella Rossellini.

But why has the salvage market blossomed in the last few decades? The precise reasons behind this boom are difficult to pin down, but there are three or four strong contenders.

The first has to be the popularity of interior design and DIY. The English in particular are a nation of nest-builders. As the social anthropologist Kate Fox has astutely pointed out, we can't help ourselves when it comes to DIY: 'Watch almost any residential street in England over a period of time, and you will notice that shortly after a For Sale sign comes down, a skip appears, to be filled with often perfectly serviceable bits of ripped-out kitchen or bathroom, along with ripped-out carpets, cupboards, fireplace-surrounds, shelves, tiles, banisters, doors and even walls and ceilings.' This obsessive territorial marking, which has reached fever pitch in recent years, provides both the supply and the demand for architectural antiques.

A second reason must be the genuine quality that comes with many old building parts. Before the days of mass production, most building materials were made by hand with enormous care and skill. This may sound like nostalgia, but most people in the trade would agree that there is nothing that can duplicate the design integrity, craftsmanship and quality of reclaimed materials. Even buildings of fairly humble origins give up high quality salvage; late Victorian

Methodist churches and chapels, for example, used good quality timber and the best craftsmen they could afford.

Many people like salvage because, despite the high prices fetched by real architectural rarities, you can still pick up a bargain. Root around any salvage yard and you'll soon find old building materials at good bulk prices: ironwork, stone slabs, hand-made bricks, roofing tiles and timbers from tree species no longer available. The same goes for Victorian roll-top baths, doors, gates and railings, radiators, pews and stained glass panels – they still represent good value and quality compared to their brand-new counterparts.

But most people love salvage because it's different. It's an easy, instant way to bring interest and beauty into your home or garden. The mixing of styles and eras, in what salvage expert Thomas J. O'Gorman calls the 'spirit of eclecticism', has become *de rigueur* in design circles around the globe. Houses that get the salvage look right effortlessly blend Victorian panel doors with 1950s diner chic or whimsical Gothic statues and bold industrial artefacts. These houses are exciting, charming and individual. So, if you're bored with the high street and the fashion for flat-pack, look no further than the world of salvage.

WHERE DOES SALVAGE COME FROM?

The short answer is, everywhere. Wherever you have buildings, you have the potential for salvage. More precisely, wherever you have buildings being refurbished or demolished, you have the potential for salvage.

As the economy fluctuates, so do the fortunes of businesses. Over the past hundred years, for example, the UK has seen a dramatic decline in its manufacturing base, leaving many factories and mills empty. Other businesses have expanded, needing newer, bigger premises in cheaper locations.

Certain types of businesses have to refurbish on a regular basis, especially those at the cutting edge of fashion or technology. Cinemas, clothes shops, hairdressers, galleries, restaurants and high-tech offices constantly update their interiors and equipment or face being branded as 'outdated'. Whatever gets thrown out has the potential to be recycled. In fact, many of the bigger salvage companies now liaise with demolition companies and architects directly.

Urban redevelopment has also proved to be a good source of salvage. In many European and American cities, old neighbourhoods are springing back to life and attracting wealthy young urbanites who want to live, work and play within the same postcode. Dockside developments are especially popular, but lots of

different types of buildings once used as factories, post offices, banks, mills or offices are being transformed into swanky apartments, leaving behind a wealth of architectural salvage.

The same goes for domestic properties. As we've seen, the British are fanatical about house 'renovation', and it's amazing to think you can still find Victorian doors, Edwardian fireplaces and 1930s bath suites on skips. This now happens less frequently than in the past, however, as home owners and builders have got savvy to the value of antique fixtures and fittings.

Other salvage comes from the nibbling away of important buildings – a gate thrown out here, a new bathroom there. In fact, many people would be surprised to find out just where some salvage comes from; for example, over the years, the London-based salvage company LASSCO has sold items from some of the city's most prestigious buildings including Buckingham Palace, the Tower of London, the Palace of Westminster, the Royal Opera House, the National Portrait Gallery, Harrods and the Natural History Museum.

Institutional buildings such as schools, prisons, parks, doctors' surgeries, law courts, swimming pools, hospitals, even the Houses of Parliament have all yielded fascinating salvage over the years. Dentist's chairs, prison loos, doctor's screens, school desks – the possibilities are endless and often the more unusual the better.

A large source of salvage is ecclesiastical buildings. The changing fortunes of the Church, especially in the UK, has meant that many places of worship have become redundant. Church attendance has fallen by about a third since the 1970s and churches often find that they are too big, in the wrong location, or unsuited to modern congregations.

As a result, some churches and chapels have been abandoned or sold off. Others have had to change their use or refurbish. Church salvage isn't to everyone's taste: some people feel uncomfortable with what they see as sacred objects ending up in mundane, domestic situations. However, I'd have to agree with Neville Griffiths of Rococo Antiques and Interiors, who points out that although buying from a church is a tricky issue, it would be worse to do nothing and let a building's treasures to be lost for ever: 'If you can reuse the materials, then you can take some of the spirit of the church and bring it back to life.'

Agriculture is another rich seam of salvage. Redundant barns and other farm buildings are highly sought after for their conversion potential and raw materials. In the US, barn salvage is particularly popular – probably due to the huge scale of these magnificent buildings, their relative antiquity and the high quality timber they yield.

DREW
PRITCHARD
&
STAINED GLASS
ARCHITECTURAL
ANTIQUES
BLAENAU-FESTINIOG

W.N.FROY & SONS LTD
COLD
HAMMERSMITH.W.
BOLDING
COLD
LONDON

WHERE TO START? VISITING A SALVAGE YARD

No two salvage yards are alike. They range from the chaotic to the highly organized. Many salvage yards simply pile up 'as found' salvage and expect you to rummage through yourself. This type of salvage hunting is hugely appealing, the act of poking around a dealer's overgrown back yard or dusty shed adding a treasure hunt thrill to the proceedings.

Other salvage yards feel like museums and focus on high-end architectural antiques, many of which have been skilfully restored by in-house craftsmen. The buildings that house these businesses are often a joy in themselves; Artefact Design and Salvage in San José, California, for example, is such an inspiring, beautiful space that its owner rents the shop for parties.

GET ACQUAINTED

Before you even think about buying a piece of salvage, it's a good idea to get acquainted with what's available. Antiques expert Alan Robertson suggests visiting three or four dealers in your area and having a browse among their stock. You will also find that some salvage yards specialize. The Directory at the back of this book will help you find a dealer near you.

TIPS FOR SALVAGE HUNTING

1. **Do your homework.** If you're planning to use salvage as part of a renovation project make sure you choose appropriate building materials and fittings. The chapters in part two, Salvage, will get you started, and Judith Miller's *Period Details Sourcebook* is an excellent reference guide for any would-be restorer. Some salvage dealers even have old retail catalogues to help you determine the appropriate fittings.

2. **Make cost comparisons.** Compare prices between various dealers. Include any restoration, installation and delivery costs.

3. **Measure up carefully.** This is important if you're looking for doors, fireplaces or anything else that has to fit an exact space. Include height,

width and depth. And don't forget to check whether your new purchase will actually fit through the door or get up the stairs. Most purchases are final.

4. **Build around it.** If you want a really special door or window you might have to build around it rather than try to make it fit an existing space. A skilled joiner should be able to tackle this fairly easily.
5. **Take your own measuring tape.** And a pad and pencil.
6. **Be prepared to rummage.** Take a strong pair of gloves. Wear sturdy shoes and clothes you don't mind getting dirty. Bring a waterproof – salvage yards are often open to the elements.
7. **Leave the kids at home.** Salvage yards aren't always safe or suitable places for very young children.
8. **Take samples and photographs.** This will help if you want to match up existing materials or designs.
9. **Calculate quantities.** Always overestimate what you'll need. If you're lucky, the yard might buy back what you don't use.
10. **If you want something specific, phone first.** Alternatively, look at the salvage yard's website to find out if it is in stock and what the price is.
11. **If you see something you really like, buy it.** If you go away to think about it, chances are when you return it will have gone.
12. **Make use of the dealer's expertise.** Many are enthusiastic about sharing their knowledge and advice.
13. **Don't be afraid to negotiate.** Be prepared to haggle on price, especially if you are buying in bulk.
14. **Get a receipt.** This will help if your item turns out to be a 'fake' or, worse still, stolen. See Provenance, page 26, for more information.
15. **Use your imagination.** Some of the most exciting houses have used salvage in new, different ways. See the Using Salvage suggestions at the end of each chapter for inspiration.

OTHER SOURCES

For the intrepid salvage hunter there are other sources of architectural antiques:

AUCTIONS

Salvage finds its way into general auctions up and down the country. You also get specialist auctions that concentrate solely on architectural antiques.

Despite the fierce competition that surrounds these sales, the average punter can still pick up a bargain. This is for a simple reason. The majority of auction lots are bought by people in the trade. People in the trade need to make a certain amount of profit on any given piece (around thirty to sixty per cent). You don't. That's why ordinary punters can afford to bid that little bit higher than someone from the trade and still come home with a piece of architectural salvage at well below the normal retail price.

There are downsides: shopping for salvage in an auction house isn't like popping into your local furniture store. It's hugely unpredictable what will turn up; you have to invest lots of time and energy into scouring the catalogues and attending the sale previews; there's no guarantee that you'll win the sale; and, unless you've done your research, you might end up buying something that will cost a fortune to restore. With a little experience and background information under your belt, however, auctions can be thrilling and a cost-effective way to fill your home with architectural treasures. Visit www.salvo.co.uk for a list of salvage auctions in your area.

THE INTERNET

The auction website eBay has opened up a whole new marketplace for buyers and sellers to trade architectural antiques. If you're selling, eBay gives you access to millions of potential customers worldwide. If you're buying, you get the opportunity to browse through thousands of antiques and collectables at the click of a mouse.

Just like real auctions, however, once you've bid for an item you are legally bound to honour that agreement should your bid be successful. You also might want to check whether the vendor will deliver, as some larger items are 'buyer collects' only. For more details, visit www.ebay.co.uk. Other Internet auction sites offer similar opportunities.

Some salvage yards are also beginning to sell their stock online. It's a great way to suss out a particular company and get a measure of their prices. If you see something you like, you can always pop into the yard or even buy it over the phone. This is ideal if you live out in the sticks or you're hunting around internationally, but dealers and customers would agree that it's always better to see what you're buying in person.

BUILDING SITES

Some people bypass salvage yards and buy straight from the demolition site. You can pick up a bargain, but this approach isn't recommended for anyone new to salvage. Although most demolition companies are entirely reputable, there are huge potential pitfalls, not least the physical dangers of a building site. However, if you're determined to have a go, here are some basic dos and don'ts.

- Don't just wander on to a building site. Speak to the gaffer and arrange a time to meet on site.
- Only buy from a demolition site if you've got something very specific in mind. The phrase 'I'm just browsing' won't go down too well.
- Take a hard hat – don't expect them to provide one.
- Expect to pay with cash. But remember that this leaves little room for comeback if you're unhappy with your purchase.
- Pay and take away the same day – otherwise you might find your purchase goes walkabout overnight or gets sold to someone else.
- Check everything: if you're buying a pallet of hand-made bricks, for example, check the middle of the stack for broken or new bricks.

OTHER POTENTIAL SOURCES

There are other sources of salvage, if you know where to look. Charity shops often have furniture; car boot sales come up with lots of surprises. Scour the small ads in your local paper; think about council recycling depots, company closure sales, farm auctions and good old-fashioned scrapyards. Even skips can be a good source of salvage, but always ask permission before you take anything.

POLYTHENE LINED
MINIMAX
FIRE EXTINGUISHER
IN CASE OF FIRE
MINIMAX LTD
FELTHAM MIDDLESEX
ENGLAND
POLYTHENE LINED
MINIMAX
CASE OF FIRE
MINIMAX LTD
MIDDLESEX
ENGLAND
FIRE EXTINGUISHER
PROTECT FROM FROST
MINIMAX LTD
FELTHAM
ENGLAND

PROVENANCE

No. It's not a region in the south of France. Put simply, provenance means place of origin or derivation. Used by the fancy antique world, it means proof of authenticity or past ownership.

But why is provenance so important? In the antiques trade, provenance separates the ordinary from the extraordinary. It gives you an artefact's journey through time: who owned it, what it was used for, and whether it was associated with anyone or anything famous. In other words, there are lots of decorative antique toilets sitting in salvage yards across the UK, but if you discovered the loo that belonged to Winston Churchill or John Lennon, that provenance raises it from the bog standard to the highly collectable.

In all parts of the antique world, including architectural salvage, provenance can add significant value to an object. Take nautical reclamation, for example. A timber from a nineteenth-century sailing vessel has value. A timber from a nineteenth-century sailing vessel involved in a famous battle would be even better. A timber from Darwin's ship, HMS *Beagle*, would be a genuine treasure (although marine archaeologists suspect that HMS *Beagle* is in fact buried deep beneath an Essex marsh so anyone who claims to have a section of the ship is probably telling a fib). But you get the point.

Establishing provenance is also important in the fight against illegally obtained architectural antiques. Antique garden ornaments, for example, are easy pickings for thieves. Despite the enormous weights involved, everything from humble pig troughs to huge marble statues are pinched with depressing regularity and sold to clients who choose not to ask too many questions. Many end up abroad, never to be seen again.

When there is high demand for something, both legitimate and black markets inevitably spring up. The popularity of architectural antiques is no different. Home owners have been known to come back from two weeks' holiday to find all their Victorian fireplaces gone. Or worshippers can arrive at their local church on a Sunday morning to discover that crucifixes, stone angels, pews and even entire altars have been lifted during the night.

Establishing the provenance of an object will not only reassure you that you are getting the genuine article; it also contributes to cracking the trade in stolen goods. Most people are totally unaware that if an item they've bought turns out to be stolen then, at best, they'll lose the goods without compensation. At worst they'll be prosecuted for receiving stolen goods.

But establishing provenance can be tricky. According to antiques expert Michael Flanigan, provenance is something often proffered but less often proved. During the Middle Ages, for example, legend has it that there were so many pieces of wood passed off as genuine bits from Christ's crucifixion cross you could have built an ark out of them.

However, the vast majority of salvage is totally bona fide, especially if you buy from dealers and salvage yards who have signed up to the Salvo Code (see page 193). These are guidelines which aim to give buyers confidence that items they buy have not been stolen or removed from listed or protected buildings without permission.

In an ideal world, an object can be traced through the sales receipts from previous owners (a bit like when you're buying a second-hand car). Other ways to establish provenance include looking at old photographs, newspaper clippings or texts that connect a person to an object.

To safeguard against receiving stolen salvage, Austin O'Driscoll, security consultant at DMG Antique Fairs, recommends that you always get a receipt for your purchase, preferably with the name and address of the dealer. It's also important to pay by cheque if you can – this leaves a paper trail of payments. If you suspect that an item is stolen, immediately contact your local police force, or if an item of yours goes missing, look at *Trace* magazine or www.trace.co.uk, a catalogue of stolen art, antiques and salvage with the corresponding police officer's details.

FAKES AND REPRODUCTIONS

There's one very important difference between fakes and reproductions. Fakes are deliberately designed to fool the buyer. Reproductions aren't.

Reproductions are common in the salvage business and exist for a number of reasons. Sometimes they're copies of original artefacts that no longer exist or are very rare. You can find eighteenth-century cast-iron copies of the priceless Medici Urn, for example, a famous marble vase that now resides in the Uffizi Gallery in Florence. Some of the older reproductions are high quality and very valuable in their own right.

Modern copies also allow a buyer to purchase an item that, if made from the original material, would be very expensive. You often find that reproduction marble statues are made from cheaper, lower quality composition marble (crushed marble mixed with cement or resin). Rather like costume jewellery, a modern composite marble statue probably won't hold its value like an original, but if you just want a bit of inexpensive 'garden bling', a modern reproduction might fit the bill.

Sometimes reproductions are used when the original would be impractical; some fantastic pieces of Victorian sanitary ware have been reproduced, for example, which have been tweaked to comply with modern plumbing requirements. Thomas Crapper & Co., for example (see Directory, page 221), sell accurate but eminently usable reproductions of their fine old designs.

But most of the time reproductions simply exist because the salvage business is a victim of its own success. Demand for certain types of architectural ornament – fireplaces, sinks, garden ornaments – is so high that it can be difficult for traders to keep up.

The problem with reproductions is that the quality can vary wildly. At the top end of the market you find excellent copies, made with genuine care and craftsmanship. Painstaking research goes into using the correct materials and construction, and this keeps old skills and knowledge alive for generations to come. Really high quality reproductions can even become collectable in their own right. Take Coalbrookdale garden benches, for example. An original antique Nasturtium cast-iron bench will fetch thousands of pounds at auction. A modern copy, made by Coalbrookdale using exactly the same casting techniques and pattern books from 1850, will cost you just over a thousand and may even become an antique of the future.

At the bottom end of the market you find low-cost, badly made reproductions, many of which are imported from the Far and Middle East where labour and raw materials are cheaper. These reproduction pieces cost a fraction of the original but are invariably of a poorer quality and finish. But however good or bad the quality of a reproduction, if the dealer has been honest that the item is a copy, you'll get what you pay for, broadly speaking. It's an entirely different matter when you've been told your purchase is an original but it turns out to be a copy. That's a fake.

Just like every other business activity, the buying and selling of architectural antiques is protected by the Trades Description Act. In other words, if you've been sold a reproduction when you were assured it was an original, you have a legal right to get your money back. In practice, this is a little more complicated and depends on where you bought your item and exactly how it was described.

If an auction catalogue or price tag describes something in writing as an original and it turns out to be a copy, you have written proof to support any attempt to recoup your money. If you bought your fake from the back of a car at the local jumble sale, chances are you'll never see your money again. Equally, keep an eye out for how something is described; as Judith Miller points out, 'There is a considerable legal difference between describing something as "Georgian" and "probably Georgian".'

SELLING SALVAGE

Most of this book concentrates on the different ways to acquire salvage. But what do you do if you have an architectural antique you want to *sell*?

SELLING TO THE TRADE

One of the simplest and quickest ways to sell an architectural antique is to call your local salvage dealer. See the Directory at the back of this book for your nearest yard. If the item is interesting enough, most dealers will be happy to take it off your hands for the right price. If it's still *in situ* (e.g. the fireplace is attached to the wall), the dealer can arrange for it to be safely dismantled and will adjust his quote accordingly.

So, how much can you expect to be paid? People are often disappointed by the amounts a dealer will offer. They're then doubly peeved when they see their antique for sale in the local salvage yard with a substantial mark-up. The reality is that dealers need to cover their running costs and make a decent profit. As a general rule, dealers add around thirty to sixty per cent to an item before it reaches the shop floor. If a piece of salvage needs restoration, the final mark-up can be much higher.

With all these different factors involved, the best way to get a fair price for your item is to keep a close eye on the market. Do your research. Visit a couple of salvage yards to get an idea of prices. Read up about the piece you're trying to sell, it will give you greater bargaining power. Be realistic about how much restoration work it will need. And don't forget to factor in delivery costs if you need the yard to pick it up.

Ask two or three dealers to give you a quote. If it's in very good condition, you might be offered about a third less than you've seen it in the showroom. If it's in worse shape, be prepared to lower your price significantly. In the end, you choose whether you accept the offer and the dealer chooses whether he can afford the deal. A happy compromise is usually found.

SELLING IN AN AUCTION

Selling an item at auction is another option. When you take your architectural antique to the auction house you'll be asked whether you want to put a reserve price on it. Unless you're absolutely desperate to get rid of something, use the reserve price option so you don't lose out if the bidding fails to take off. If you're not sure how much your reserve price should be, the auction house will be glad to advise. Remember, it's in their interest as well as yours to make a good profit.

The downside to selling at auction is the various costs involved. You'll have to pay a seller's commission: ten to fifteen per cent of the final selling price. You'll also have to factor in the cost of a photograph in the auction catalogue (if you want one), transport costs and insurance. Most of these costs are subject to VAT. It's also worth bearing in mind that you should avoid putting anything into a local auction that has been rejected by local dealers. Chances are, the same dealers will be at the auction and won't bid for your item. If this happens, consider using an auction house outside your local area or selling via the Internet or classifieds.

Internet auctions work on the same principle as normal auctions. You set your reserve price (if you want one), sit back and hope that two or more people start bidding. The most popular site, eBay, will give you access to a huge number of potential buyers. If you want to sell salvage this way, there are three golden rules: grab people's attention – choose a catchy title that describes your item in one sentence; describe your item as fully and honestly as possible; and always include at least one picture, which will greatly increase the amount of bids you'll get.

OTHER MARKETS

If you've got more than one item to sell, you might want to consider setting up a stall at a local car boot sale. Dealers keep an eye on these and you can shift most items if you've priced them fairly. Classified ads are another possibility – magazines such as *Period Living & Traditional Homes* have a section at the back where you can advertise. And don't forget local antique fairs.

RESTORATION

To restore or not to restore? Difficult question. Most people in the salvage trade would advise real caution to anyone thinking about restoring an architectural antique themselves. The area is fraught with danger and, unless you are very skilled, you could end up knocking hundreds or even thousands of pounds off the value of an object.

One common mistake, especially with architectural antiques, is to over-clean an item and ruin its patina. Patina is the lovely lustre an object acquires as it ages – a combination of grime, polish, wear and tear, and the effect of daylight. Much of the value of an antique depends on its patina.

Take bronze, for example. Bronze develops a gorgeous sea green patina when left outside and its presence greatly increases the value of a statue or urn. People love this effect so much that for centuries forgers have been inventing sneaky ways to create the same effect artificially.

Most people who buy salvage like to see the history and romance of an object. This, as often as not, is reflected in its cracks, tears, dents and stains. Wooden shutters with their peeling layers of paint, an urn covered in moss and lichen, iron railings with a bit of rust here and there – it's the opposite of the sanitized, uniform, freshly painted world of modern interior design.

However, there is a difference between a charming, well-worn piece of salvage and one that's totally knackered. Many pieces at salvage yards do require some restoration: old radiators rusted on the inside, roll-top baths in dire need of re-enamelling, incomplete cast-iron fireplaces and so on. Making a poor repair to an architectural antique will lower its value so, when you do have a salvage SOS, who's the first person to call?

IN-HOUSE RESTORERS

Some of the larger salvage companies have their own in-house restorers. If you see a piece of salvage you like, unrestored, ask for the price restored. Even with the extra cost involved, many pieces of salvage are still great value. Even if the salvage yard doesn't do its own restoration, the owner may be able to put you in touch with someone in the local area who could help.

The same applies if you need a sympathetic tradesman to install your new piece of architectural reclamation: most salvage yards have been in existence

long enough to have built up a good network of 'salvage-sympathetic' plumbers, builders and carpenters.

The Directory at the back of this book can help you find a restorer, but there are three other useful sources of information for those seeking reputable restorers for their salvage:

THE CONSERVATION REGISTER The Conservation Register was first set up in 1988 by the Museums and Galleries Commission as a database of restorers across the UK and Ireland. It's now operated by the United Kingdom Institute for Conservation (UKIC) in partnership with Historic Scotland and the National Council for Conservation-Restoration. Visit www.conservationregister.com (or see Useful Addresses, page 229) and you can search for restoration and conservation experts in all sorts of relevant fields including architectural carving, bronze sculpture, wrought iron, terracotta, woodwork, stained glass and stonework.

PERIOD PROPERTY UK Period Property UK was established to help people share their passion and knowledge about living in and restoring old buildings. The website www.periodproperty.com has a very useful 'seeking specialists' section, a list of suppliers, services and craftsmen across the UK who work in various areas of restoration.

SALVO The Salvo website www.salvo.co.uk has a directory that you can search, by area, for restorers and consultants around the world. The links section also contains names, addresses and websites of salvage restoration experts and craftspeople.

TIPS FOR CREATING THE SALVAGE LOOK

It's difficult to define the 'salvage look'. Reclaimed interiors are different every time. Unique and adventurous. Fantastical and eccentric. Elegant and timeless. The antithesis of flat-pack fashion, salvage attracts people who want a home that's original, individual and truly personal to them.

No two salvage interiors are the same. And how can they be when reclaimed architecture comes in such an infinite variety of forms – statues from a French chateau, Art Deco office chairs, marble bath-tubs from an English stately home, angels from a Scottish chapel, a 1950s neon sign from an American motel – the possibilities for innovation are endless.

The tips below are designed to give you a head start, but the rule is that there is no rule. Make it up as you go along. Experiment. Create. Chop and change. Above all, enjoy yourself.

1. **Resist the urge to over-restore.** Battered floors, peeling paint, weathered stone – these all have an intrinsic beauty and subtlety that reflect their history, character and previous use. In many cases, wear and tear of this kind actually adds value to a piece, hence the brisk trade in artificially aged reproductions.

2. **Mix up the time periods.** Revel in a spirit of eclecticism. Don't be afraid to put a Greek-style column next to a 1940s cinema seat, or a Victorian butcher's block in a modern stainless-steel kitchen. Different architectural styles will bounce off each other, enhancing the idiosyncrasies of each.

3. **Find inspiration.** Visit old stately homes. Buy yourself an armful of interior magazines. Borrow architecture books from your local library. These will deepen your knowledge and give you bright ideas for design schemes.

4. **Play around with scale.** Experiment with outsized pieces of architectural reclamation in your home or garden; some of the most dramatic and exciting interiors focus on just one stunning piece of salvage – a huge marble urn or an exquisite panel of stained glass, for example.

5. **Forget minimalism.** The salvage look is about old-fashioned comfort: natural materials, layers, colours and textures. Surround yourself with objects that inspire you, that reflect who you are. Choose objects that feel homely, lived-in and well loved. Collect and clutter. This isn't a look for people who want stark, bare surroundings.

6. **Avoid sticking to house salvage.** Church pews, shop signs, school radiators, wooden factory benches – commercial and industrial salvage like this can also look fantastic in the domestic setting.

7. **Don't be squeamish.** Some of the most interesting salvage comes from places you might not expect – or like. Hospitals, dentists, sewage works, chemists, butcher's shops: some people feel uneasy about salvage from these places, but they're missing out on fantastic bargains. Hospital operating

DOMESTIC
BAZAAR

theatres often have trough-like porcelain sluices, for example, which when cleaned up make large, elegant kitchen sinks.

8. **Bring the outside in.** And vice versa. Antique statues, pretty wrought ironwork or railway benches all look stylish when brought indoors. Equally, as long as the material is suitable, indoor salvage can be put to good use outside. Stone columns, porcelain sinks and coloured window panels, for example, all work in a garden.

9. **Move objects around.** A piece of salvage, once bought, doesn't have to stay in the same place. A statue in the garden will look great during summer. Then, come winter, bring it inside and pop it in the bathroom for an eye-catching focal point.

10. **Think creatively.** Change an object's function. Glass light fittings and chimney pots make fantastic planters. Doors turned on their sides become handsome panelling. Victorian bath taps make funky coat hooks. From the simple (turning a bible stand into a cookbook stand), to the extraordinary (transforming an oak confessional box into a wardrobe), your imagination is the only limit to what can be done with salvage.

COMMERCIAL APPLICATIONS

Salvage isn't just for home owners. Shops, restaurants and businesses are also getting in on the act.

Take British pubs, for example. It's not long ago that breweries were ripping out original panelling, oak beams and brass fittings to replace them with modern materials such as plastic and fibreboard. Nowadays the trend has reversed and many pubs, realizing the demand for authentic interiors, are attempting to restore these 'olde worlde' features. The most successful of these conversions have been ones that don't create a pastiche of the past – what might be disparagingly called 'horse-brass chic' – but ones that have taken the time to source original fixtures and fittings.

People such as Andy Thornton were quick to pick up on this trend. He started his business twenty-five years ago shipping salvage to the US and selling it to the nostalgia-driven hospitality industry. British breweries soon followed the American example, and before long Andy was employing a team of people to design and build pub, hotel and restaurant interiors. Today, Andy Thornton

NIGELLA LAWSON HOW TO EAT
Lyric

Architectural Antiques offers the world's largest selection of architectural antiques and employs over two hundred people.

And it's not just pubs that use architectural antiques. Rose & Co. apothecary in Leeds is a fine example of a salvaged shop interior. Open the door and you instantly step back in time to an old apothecary. The shop is packed from floor to ceiling with beautiful, mahogany glass display cases stuffed with antique packaging and perfumery, while enamelled signs advertise salves, ointments and remedies from the past.

Just down the road from Rose & Co., you'll find Georgetown, a restaurant housed in the old Dysons' Jewellers building of 1865. Rather than rip out all the original shop fittings, this enterprising eatery has made the most of its architectural antiques. The restaurant has been built to incorporate the old jeweller's signs, clocks and cabinets, making it one of the most unusual and exciting architectural spaces in the city and winner of the 1994 City of Leeds Annual Design Award.

In London, the *Silver Sturgeon* river cruiser has used architectural salvage to great effect. Antiques from auctions in London and Paris re-create the atmosphere of a 1930s transatlantic sailing vessel, and include burnished metal grilles from Regis House, lights from the Palais de Justice in Marseilles, and nineteenth-century deer guards from a French chateau.

But you don't have to spend a fortune to add a touch of salvage to your business. Foyles Bookshop in London, for example, commissioned Retrouvius Reclamation & Design to design and create benches, tables and stools made of reclaimed timber gleaned from local demolition sites.

Old beach huts, kiosks and confession boxes make great changing rooms. Chapel chairs create comfortable, inexpensive café seating. Pews are ideal for pubs. Two pedestals and one door make an instant beer garden table. Old exterior hanging shop signs and enamel plates significantly add to the appeal and display potential of a shop. Advertising ephemera, neon signs, old vending machines, bottles and original packaging – these all add to the retail ambience.

SALVAGE

BATHROOMS

Bathrooms are a fairly new invention. Before the mid 1800s most people's idea of a toilet was an earth closet in the garden, effectively just a hole in the ground, and a chamber pot if you were caught short during the night. Washing was also a simple affair, just a jug of water and a basin in your bedroom or dressing room. Baths were a rare treat. Buckets of water had to be heated on the stove or fire, lugged across the house and tipped into a portable metal bath. When bath time was over, it wasn't a just a matter of pulling the plug – every last drop was emptied by hand.

By the middle of the nineteenth century this traditional system was no longer practicable. The Industrial Revolution was well under way and people from all over Britain had flocked to the towns and cities in search of work. Overcrowding was appalling and London, like all the other major centres, suffered from fouled fresh water supplies and overflowing sewage. Disease was rife: in the 1830s the death rate among under-fives was almost fifty per cent in towns. Typhoid, cholera, diarrhoea and dysentery were common causes of mortality and almost exclusively caused by poor sanitation. Even the rich weren't immune to the problems; in 1861 Queen Victoria's consort Prince Albert died from typhoid. Something had to be done.

The Victorians, ever ingenious, threw themselves into the task of creating water and sanitation facilities on an enormous scale. Huge sewage systems were built, pressured water began to be piped into people's houses, and new ingenious heating methods were devised. People could finally control water coming in and out of their homes. These great leaps in engineering also inspired the production of bathroom fittings: indoor flushing toilets or 'water closets', mass-produced baths and ceramic hand basins. For practical reasons, it made sense to have all these items plumbed near each other in one distinct room. The modern concept of the bathroom was born.

At first, the general public were cautious about these newfangled ideas. The notion of going to the toilet in one's house was too much for many Victorians, who thought the idea wholly unhygienic. To cope with the transition, many fittings from early bathrooms were encased in wood panelling, their embarrassing function hidden from view. Sinks were disguised as dressers, baths became large chests, and toilets were often hidden behind a curtain or screen. Many of these wooden boxed fittings are hugely collectable today.

By the end of the nineteenth century, however, people had got used to the idea. The wood encasements gave way to sleek, easy-to-clean porcelain fixtures with ornate designs and flowing curves. New homes included an indoor toilet, bath and washbasin as a matter of course.

TOILETS

One of the greatest achievements of the Victorian age was the development and refinement of the 'water closet' or WC. The idea of a flushing lavatory, however, is not as new as you'd think. As long as four thousand years ago, King Minos's palace in Crete and the Indus Valley civilization (now Pakistan) were enjoying the luxury of flushing toilets. Centuries later, the Roman Empire insisted on flushing latrines in many of its buildings and, even in the Middle Ages, fine examples of flushing loos can be found. In the twelfth century, for example, the Abbot of St Albans built his own rainwater cistern and was probably the first Englishman ever to own a water closet.

However, the modern toilet as we know it was finally perfected by the Victorians at the end of the 1800s and has changed little since. It basically consists of a ceramic toilet pan, a gas trap and a cistern.

Ceramic pan designs have remained fairly constant over the years – white porcelain is still standard for most bathrooms. The Victorians loved ornate patterns and decorated some of their white toilet pans with blue, red or pink transfer prints. These items now fetch high prices at auction and salvage yards. More recently, the 1970s saw a fashion for one-coloured bathroom suites, avocado, turquoise and beige being popular choices. Unfashionable nowadays, these suites are even considered to bring down the value of a house. But, who knows, they might just be the salvage of the future. Then again . . .

401

Before the Second World War most toilet pans were made from biscuit-ware, clay that was fired, glazed and then fired again; this was an expensive process but it did produce sharp lines, elegant edges and gave greater definition to any surface decoration. After the Second World War most lavatory pans changed to vitreous china, a material that only needs to be fired once but produces more rounded, softer edges. When buying reclaimed sanitary ware, the age of a piece will be given away by the crispness of the edges and decoration.

Cistern design has moved on a little, from the traditional wall-mounted model with a chain pull to the modern close-coupled cistern that sits on the back of the toilet pan. Old-fashioned wall-mounted cisterns relied on gravity to flush the water down the loo. These were the main type of cistern well into the twentieth century and were mostly made from cast iron, wood lined with metal, or ceramic. Modern close-coupled cisterns use a more complex siphon system, which means you don't need to hang the cistern high on the wall.

Reproduction toilets often mix Victorian- or Edwardian-style ceramic toilet pans with close-coupled cisterns – this combination would not have existed in history but can offer the customer a way of putting a period-style toilet under a window or in a bathroom with a low ceiling.

BATHS AND SHOWERS

As we've already learned, having a bath before the 1850s would have been time consuming and sporadic. Bathing took place in portable metal baths – often tin or copper – which had to be filled and emptied by hand. To keep the bather comfortable, the bath would be placed in front of the fire, but the heat-conducting properties of metal meant that it was difficult to keep the bath-water warm for long. After the introduction of mains water, plumbing and sewage disposal in the mid nineteenth century, the act of having a bath was transformed. The bath went upstairs to join the new toilet and washbasin and the bathroom started to become a place for private indulgence and beautifying treatments. Nowadays we view the bathroom as one of the most important rooms in the house. It's often still the smallest room in any home, and yet we spend thousands on fixtures, fittings and decorations.

Despite the early phase of disguising baths with wooden panelling, by the end of the nineteenth century roll-top cast-iron enamelled baths were the tub of choice for most homes. White vitreous enamel on the inside, the outside of the bath could be decorated to fit in with the rest of the bathroom – using paint

SHOWER
BATH

effects such as marbling or stencilling – while the feet were modelled on extravagant classical designs such as ball-and-claw feet and mythical beasts. During this time, baths were also still being produced from sheet lead, copper or zinc. These are very popular in today's market and look magnificent restored.

In contrast to baths, at the turn of the century showers were less common and mostly restricted to wealthier households with sufficiently powerful plumbing. Earlier in the 1800s there had been a few gallant attempts to produce a separate shower cubicle – the English Regency Shower being a prime example – but even by the end of the 1800s most showers were only 'shower sets', a hand-held shower head attached by a flexible pipe to the bath taps. On the other side of the Atlantic, things were little better. If you're lucky enough to see one, the American Virginia Stool Shower is a joy to behold. This fantastic contraption, designed for a sit-down shower, was developed in the 1830s. The wooden unit was essentially a revolving seat, like a piano stool, with a hand-operated lever that pumped water directly at the bather's face. The *pièce de résistance* was a foot pedal that controlled a stiff scrubbing brush which worked its way up and down your back, so you were blinded, flayed and soaked all in one go.

By the early twentieth century sophisticated showers with a fixed overhead douche were increasingly popular and, only a few decades later, large hub-cap style shower heads came to epitomize 1930s five-star glamour and decadence.

WASHBASINS

Prior to the introduction of the bathroom, the necessities of washing would be carried out using a jug of water and a basin. These would sit upon a washstand – a wooden table or cupboard often marble-topped to prevent water damage.

With the advent of hot and cold running water in the mid nineteenth century, plumbed-in washstands were fitted with an inset basin, plug and taps, but their wooden furniture styling remained essentially the same. It was only with the introduction of ceramic sinks, towards the end of the nineteenth century, that the modern bathroom sink began to take shape. Sometimes screwed to the wall, sometimes supported on a ceramic column, the earliest examples are often highly decorated. By the turn of the century, however, white had become the standard colour and ceramic sinks have changed little in design since that time.

As with Victorian toilets, the more decorative the ceramic, the more valuable they tend to be. Look out for transfer-printed examples, fancy pedestals or columns, or decorative washstands made from wood or cast iron.

FINISHING TOUCHES

One thing you won't find a shortage of in a salvage yard is taps. The designs may have changed since Victorian times, but the technology is essentially the same. The market for modern reproduction taps is strong, but with a little TLC reclaimed antique taps can add real sparkle to a bathroom. Equally, modern-looking taps on a reclaimed bath will completely spoil the effect.

Most salvaged taps will be made from nickel, brass or chrome. Reclaimed brass taps, in particular, are always good value and of good quality compared to their modern counterparts. The same goes for other bathroom accessories, such as heated towel rails, toilet roll holders and metal bath racks.

SOURCES

Most salvage yards have the odd piece of bathroom salvage. Many have a wonderful selection to choose from and will help you put together an entire bathroom suite – from toilet to bath taps – fully restored, fitted and plumbed. Some dealers also offer reproduction bathrooms and fittings. Depending on the quality of the reproduction, this can be a quick and easy way to get a period look or replace a part that you've been struggling to find. Some companies, such as Thomas Crapper & Co., sell accurate copies of their old models.

You can almost pay as much as you want for bathroom salvage. The cheapest sanitary ware is well priced and of a high quality compared to new bathrooms, but the exceptional pieces can be very expensive indeed. Canopy baths, wooden framed toilets (thunderboxes) and highly decorated ceramic basins and pans – these all fetch high prices in salvage yards, but create unforgettable centrepieces in any bathroom.

The first thing to consider is what the bathroom will be used for. Cast-iron roll-top baths make durable and attractive family baths, but a delicately decorated Victorian basin may be better placed in an occasionally used cloakroom. You should also think about the proportions of the room. Modern bathrooms are often quite small and antique ware can be imposing. Make sure you've measured the space properly and that the floor can take the extra weight (a cast-iron bath full of water is enormously heavy).

If you want to hunt for bathroom salvage yourself, you won't have to look too far before you find a farmer's field containing a roll-top bath or a skip with a 1930s sink. Bathrooms are one of the most commonly refurbished rooms in the house, so there's never a shortage of older fittings if you know where to look. In fact, the problem is not the availability of bathroom salvage, but how much a piece will cost to repair and install.

RESTORATION AND CLEANING

The restoration of working fittings like taps and flushes is a tricky job. Unless you're a very skilled DIYer, it's probably best left to a professional restorer who will know where to get replacement parts such as washers and seals. In fact, most people find the best option is to buy a salvaged bathroom from someone who has already done all the hard work. Make sure the company delivers and fits your bathroom too; you don't want to pay a fortune for a restored piece, only for it to be chipped or scratched in transit.

You can tackle some jobs yourself, however. To smarten up a discoloured roll-top bath, apply a whole bottle of cream cleaner and leave for twenty-four hours. Failing that, lime scale remover and household bleach should take off any surface grime, while a combination of ammonia and peroxide on a cotton wool pad will draw out dirt from cracks – especially if you leave the pad in place, cover it with clingfilm and leave overnight. If the enamel in a bath is too damaged or discoloured to repair, don't worry: it's relatively easy and cheap to get it re-enamelled by a professional.

As for the outside of a roll-top bath, a coat of paint will give it a new lease of life – just make sure that your bath was designed to be painted in the first place. You might want to avoid enamelled baths that are very rusty or have missing feet, however, as the cost of restoration can be prohibitively expensive.

Other things to bear in mind:

- Look for a maker's name: manufacturers such as Royal Doulton, Twyford and Shanks are always a safe bet.
- Antique traps are often low to the floor – you may need to cut a hole in the floorboards or raise the bath.
- Make sure your bath has an overflow to meet with modern plumbing regulations.
- Is your boiler big enough to fill a Victorian bath?
- Budget for piping and brackets to suit the bathroom, otherwise the look will be spoiled.
- Make sure ceramic basins don't have any cracks in them.

USING BATHROOM SALVAGE

The main reason to buy reclaimed sanitary ware is to create an atmosphere that simply can't be evoked with reproduction pieces. If you're renovating a period bathroom, it's essential that you at least try to obtain original fixtures and fittings. It's also important to get the details right: make sure the taps, plug, overflow and shower fittings are the same material and style.

Victorian and Edwardian bathroom fittings can also work on their own, as individual pieces. A beautiful copper bath or an extravagant cast-iron roll-top bath will look magical in the centre of an otherwise stark white bathroom. A really ornate lavatory, on the other hand, is so attractive and rare that it might warrant a special little room of its own.

Kooky pieces of sanitary ware create a real talking point. Victorian prison loos, which are often made from metal, look strangely at home in a contemporary setting. Bathroom fittings from hotels, schools, boats, ocean liners and trains have bags of character and there's often a great story behind them.

If a piece of bathroom salvage is beyond repair, consider other uses for it. Hot and cold taps mounted on a plank of wood as a quirky coat hook; ball-and-claw feet wall sconces; bathroom tile coasters; and washbasin pedestals as supports for tables, sundials and birdbaths.

EXPERT TOP TIPS

Sam Coster of Mongers: Reclaimed and antique sanitary ware

- If there is crazing to the glaze of earthenware WC pans and cisterns, check that they can hold water.
- Check very carefully for cracks in basins, particularly if they have been stored outside – frost can do a lot of damage to porcelain.
- Unless you wish to keep the taps, it is better to buy a basin with taps removed as it is easy to break the basin removing them.
- Check that taps have been taken apart before re-plating.
- Bath resurfacing is only as good as the person doing it. Find someone with a good reputation who has plenty of experience.

BRICKS

Brick has been around since at least the tenth millennium BC. The ancient civilizations of Mesopotamia and Egypt made good use of both sun-dried bricks (also known as adobe) and fired bricks in all sorts of buildings from houses and river walls to the enormous pyramids. Sun-dried bricks were simply that – blocks of clay left to dry in the sunshine – but the process of firing natural clay turned the mud-like substance into a hard, durable and inert building material.

Bricks and brick building arrived in Britain with the Romans, who set up brick and tile factories on the edge of many settlements. The Romans were skilled brickmakers and their fired clay ended up lasting much longer than their colonization of Britain. It's not uncommon to find reused Roman bricks in many ancient structures around the country: the Anglo-Saxon builders of the Church of the Holy Trinity in Colchester, for example, built the tower to include a large proportion of reclaimed Roman bricks.

After the Romans left in the middle of the fifth century, the art of brickmaking largely disappeared in Britain. It was only in the thirteenth and fourteenth centuries that the technology and knowledge returned. The brickmaking craft had been kept alive in Italy and Byzantium during the intervening centuries and, as a result of increasing trade during this period, spread slowly back across Europe.

Roman bricks are large, thin and square; the new medieval bricks were easier to handle, smaller and similar to the modern brick. They were, however, very expensive to produce and still considered a prestige building material. It's no wonder then that some of the wealthiest people of the day chose brick as the material of choice for their grand houses – Herstmonceux Castle in Sussex being one of the finest examples still standing.

It wasn't until the Industrial Revolution that hand-made bricks gave way to mechanized mass production. Better kiln technology, the development of machines that could make bricks, and the discovery of new types of clay as a result of deeper mining, allowed the Victorians to produce uniform, reliable and inexpensive bricks on a truly enormous scale.

Today the brickmaking industry is nowhere near as large as in the Industrial Revolution – a fall in demand, changes in construction methods and the escalating costs all contributing to its reduction – but the brick still remains one of the most popular and prevalent building materials in the country. Reclaimed hand-made bricks are especially popular. Used by people who want to restore an old building using authentic materials or simply like their individual, hand-made appeal, reclaimed bricks have become a much sought-after commodity. There's also a strong ecological argument for reusing bricks: their manufacture takes a lot of resources both in terms of raw materials and energy.

However, buying and using salvaged bricks isn't a straightforward exercise and, before you start, you need to know what type of brick you're looking for and where it's going to be used.

TECHNICAL CONSIDERATIONS

Until the middle of the twentieth century most bricks were made and used within a thirty-mile radius. Transporting heavy bricks across the country didn't make good economic sense, especially since every region had the raw materials (i.e. clay) and labour on its doorstep. Some of the larger buildings and civil engineering projects even had their own brickworks erected on site.

Because of the different physical properties of clay in each area of the country, along with regional methods of moulding, firing and finishing, bricks differ wildly across Britain. Cambridgeshire bricks are a creamy pale beige, for example, while Staffordshire engineering bricks are a grey blue. If you're buying bricks for a conservation project it's vital to get the right bricks for your region. Do your research before you visit a salvage yard.

Make sure your bricks have been sorted on the basis of durability – some are not suitable for the outside walls of a building. The inner walls of a house can be built using common bricks that don't need to be frost-resistant. The outer walls will need to have bricks that are moderately frost-resistant, and some areas that are constantly saturated with water, such as windowsills and copings, will need to have totally frost-resistant bricks. Consult a builder about this.

For outside paving you'll definitely need frost-resistant bricks. Look for reclaimed pavers, because reclaimed wall bricks are not always suitable. You can, however, use non-frost-resistant bricks for indoor flooring, chimneybreasts and any other internal brickwork.

SOURCES

Salvage yards that specialize in reclaimed building materials will have a good selection of local bricks. You may find they range in age from rare medieval bricks to modern wirecut bricks (a predecessor of the machine-made brick).

How do you date a brick? Although the thickness has changed over the centuries, it's not always a reliable marker of age: very large bricks, for example, have been used in East Anglia for centuries. However, as a very general rule, modern bricks are 65mm thick. Late seventeenth- and eighteenth-century bricks are between 57 and 65mm thick and bricks made in the mid seventeenth century and before are about 50mm. The quality can vary enormously, and it's not true that early bricks are necessarily less durable than more modern examples. You don't have to look too far to see shoddy nineteenth-century brickwork in cheap industrial terrace housing, while many two thousand-year-old Roman bricks are still going strong today.

Reclaimed bricks aren't cheap, but they're worth every penny. Their high cost is a reflection of demand, but also the huge amount of work that goes into salvaging bricks from a demolition site. Old bricks need to be selected, cleaned by hand, sorted and stockpiled. Expect to spend a reasonable amount of money on reclaimed bricks and go for the highest quality you can afford. See the batch of bricks you are buying, or at least a representative sample – say two to five thousand bricks – to make sure you know what you're getting. Make sure the bricks are squarely and neatly packed.

If you can't find reclaimed bricks for your project, you can still buy bricks made in the traditional way by hand and wood-fired. These are expensive but remain a good option for people looking for bricks with unique texture and colouration. There's also no danger of your supply running out.

Failing that, you can buy simulated reclaimed bricks which have been distressed in manufacture to emulate an old brick. The upside of these bricks is that they conform to modern building standards in terms of strength, durability, frost resistance and so on. The downside is that they rarely have the mellow quality or intrinsic beauty of a genuine reclaimed hand-made brick.

RESTORATION AND CLEANING

Old bricks can be fragile, so you have to clean them with some care. A brick hammer is probably the best tool for removing old mortar. Don't put your brick on a hard surface as you clean it, because you'll find the brick will chip or crack easily. Hold the brick in one hand and strike off the mortar with the other. If the brick was mortared with lime, the mortar should come off fairly easily. If a more recent cement mortar was used, it's a much more difficult task.

USING RECLAIMED BRICKS

If you're going to use reclaimed bricks, make sure you buy a sufficient quantity for the whole work, because it will be very difficult to find matching ones at a later date. Work on a ten per cent wastage factor (in other words, buy ten per cent more bricks than you need).

Even the best and most expensive reclaimed bricks will not look good if used by an unsympathetic builder. Choose a contractor who is knowledgeable about reclaimed materials. Reclaimed bricks should ideally be laid with traditional lime mortar. Before the twentieth century all brickwork was laid with lime mortar, so if you want an authentic-looking wall, this is by far the best material to use. It's also easier to clean off than cement mortar, should anyone want to reclaim the bricks in the future.

EXPERT TOP TIPS

Richard Parrott of Cawarden Brick Co.: Bricks

- Always make sure you have a representative sample of the brick that you need.
- Accurate measurements are essential – it must be an average of all the bricks, as hand-made bricks vary in size.
- Go and view the bricks personally; don't rely on a description or photograph.
- Always see the whole pallet – the rubbish bricks will be hidden in the middle.
- Allow plenty of time to source the bricks, as they are often very difficult to match.

CHURCH SALVAGE AND COLUMNS

A large proportion of salvage comes from ecclesiastical buildings. During the twentieth century around ten thousand churches and chapels were sold off by the Church of England, Catholic and Methodist authorities, and approximately seventy churches are still put on the market every year in Britain. The buildings often become housing: large city churches get turned into flats while smaller, rural churches frequently change into individual family homes. During this process, the contents of the redundant churches often find their way to a salvage yard or auction house.

If a church is demolished, it can be for a number of reasons. Since churches are generally considered historically important buildings and worthy of preservation, salvage often comes from those that have suffered serious damage from fire or flood, have been neglected beyond repair, or have been closed due to the dwindling of congregation numbers.

Declining attendance may be a significant factor behind churches and chapels getting rid of their fixtures and fittings, but it's not the complete picture. The Church is like any other business: it constantly modernizes itself. Churches that are thriving often sell off their contents as part of a renovation project or to accommodate new congregations. Old churches often don't want to spend money repairing and restoring some of their antique fittings, preferring instead to swap old for new. Other churches have become multifunctional buildings: fixed pews get taken out and replaced with movable chairs. It's estimated that half to three-quarters of all church salvage comes not from the demolition of ecclesiastical buildings, but from their refurbishment. The same goes for cathedrals, which employ craftsmen and restorers to maintain the fabric of the

building; it's not unusual for worn-out, unwanted ornamental stonework, pews and other wooden items to be sold on to dealers.

There's a wide range of ecclesiastical salvage on offer, including stained glass, statues, pews, altars, chapel chairs, crosses, doors, textiles, fonts, lecterns, lighting, candlesticks, organs, tables, panelling, fonts, prayer desks and pulpits; not to mention all the usual reclaimed building materials such as stone, cast iron, bricks, tiles, timber and marble, if the building has been demolished. Columns are also a common feature of church salvage (although they can come from a variety of buildings) and deserve a special mention in themselves.

When we think of columns we think of the ancient Greeks, who devised three different types or 'orders', called Doric, Ionic and Corinthian. A column is made from three components: the base, the shaft and the top (capital). The Doric column was the simplest of them all, an elegant, stocky, plain column with flutes on the shaft and no decoration on the base or the capital. The Ionic column has more flutes on the shaft and lovely curling scrolls on its capital. The Corinthian order, which came later than both the Doric and Ionic column, is the most elaborate of the three. Its capital is carved with stylized acanthus leaves and the shaft has twenty-four sharp-edged flutes. The Romans took the classic Greek orders of Doric, Ionic and Corinthian, copied them in the first century BC and then developed their own Tuscan and Composite columns, the latter being an extravagant amalgamation of all the others.

After the fall of Rome, columns didn't enjoy a revival until the fourteenth and fifteenth centuries, when the Renaissance rekindled the western world's romance with all things classical. The impact of the Renaissance on future architecture was enormous, informing different styles of building from early eighteenth-century Palladianism, with its huge Roman columns, to the mock-classical suburban executive homes of the late twentieth century.

SOURCES

Some salvage yards specialize in ecclesiastical salvage, but many general salvage yards also have a good selection. The majority of salvage will probably come from the Victorian era, when there was a programme of church building on a scale not seen in England since the twelfth century. The churches of this era contain a wealth of good quality salvage, often made by master craftsmen employed by churches that had benefited from the Act for Promoting the Building of Additional Churches in Populous Parishes, passed in 1818, which made

£1 million of government money available for new churches. Before the act, many churches were failing to meet the needs of their poorest parishioners. Rapidly expanding cities such as Birmingham and London struggled to keep up with demand for church places, but the money from the act helped to fund many of the high quality nineteenth-century church fittings we see today.

Churches throughout history have also been great patrons of local craftsmen and artists. The Church was often the only establishment, apart from landowners, wealthy enough to procure the services of stained glass makers, master carpenters and metalworkers, so you often find a very high standard of architectural fixtures and fittings in ecclesiastical buildings. A large amount of work from the famous William Morris & Co., for example, ended up in churches up and down the UK.

As well as salvage yards, you could also look at specialist auctions. These often deal with the higher end of ecclesiastical antiques, but are well worth a visit. If you want to bypass the salvage yards and auctions, and buy from another source, be careful not to end up with stolen salvage. Churches have been hit badly by thieves stealing architectural antiques, often taking advantage of the fact that churches remain unlocked for a large part of the day to encourage people to visit. Buying from a company that subscribes to the Salvo Code will help guard against such problems.

Another possible source of salvage is the maintenance departments of large churches and cathedrals: as part of routine repair and renovation work old fixtures and fittings are sometimes removed and available for sale. It certainly can't hurt to ask.

USING CHURCH SALVAGE AND COLUMNS

For some people, buying church salvage is a sensitive issue. If you don't feel comfortable about it, that's fine. But what it is important to recognize is that many ecclesiastical architectural treasures were, in the past, just chucked on the rubbish heap. There's not only a good pragmatic reason to reuse old church salvage, but there's a strong emotional reason: it would be a travesty to see such treasures thrown away. One thing to remember, however, is that some pieces will have religious significance: use church salvage, especially statues and imagery, with respect for others.

Until recently, most church salvage ended up in the commercial sector, used to decorate pubs or shops. A large proportion still ends up in this area, but home owners have also started to see the potential in architectural antiques from religious buildings. Ironically, some churches are also beginning to buy back ecclesiastical salvage.

Pews are probably the most obvious form of church salvage: beautiful to look at, well made and easy to transfer into a contemporary setting. Use a full-length pew as a garden bench or occasional seating in a hallway. Even better, buy two pews and put one either side of a farmhouse table in your kitchen; seven-foot pews can seat four people, so they make ideal family seating. If you run a shop, pub, playgroup or café, church pews also make cheap, durable and attractive seating for customers.

If a pew is too long for a particular room – cut it down. Make a short bench or a chair and use the leftover horizontal seat timber for replacement stair treads, shelving or a window seat. You can use the back of a pew as wall panelling or a tall, elegant cupboard door. And don't forget the ends of pews are also useful too: they make great Gothic candle shelves.

Wooden chapel chairs are a real bargain. They're usually well made, well proportioned and still represent good value for money at between £5 and £30 each. Pick up a set of four or six and ask for a discount. One of the great bonuses of chapel chairs is that each one has a prayer book shelf on the back – ideal for magazines, books or whatever you might want to read at the table.

If you want the Gothic look in your home, church salvage is for you. Most Victorian church design was inspired by Gothic architecture which, with its pointed arches, stained glass, heraldic emblems and gargoyles, had been around since the Middle Ages. Look out for imposing stone or carved wood fireplaces, chandeliers in wrought or cast iron, tapestries, large flagstones, old church furniture (especially robust pieces such as oak dining tables), and arched screens. Church salvage also looks good in an Arts and Crafts setting; keep an eye out for hand-made pieces of church furniture, especially with cut-outs and carving, dark wooden floorboards, wood panelling, anything with a carved motto, church tiles in intricate designs and colours, stained glass and wall sconces.

There are also quirky ways to use church salvage: an oak confessional box can be transformed into a wardrobe, telephone booth or shower. Bible stands make great dictionary or cookbook stands. Slide holiday snaps into a hymn board or use it as a message board. For stained glass ideas, see the Windows, Stained Glass and Shutters chapter (page 164), while the Garden Salvage section (page 116) will give you inspiration for ecclesiastical stone statues.

Exterior and interior doors, leaded glass screens, wooden panelling, floorboards – all these valuable architectural components from a church are frequently cheaper as salvage than bought new. The proportions of church furniture and fittings can be a bit overwhelming, but you can either get them altered or embrace their outsized charm. Church salvage always looks good in loft apartments, where you can play around with scale and proportion.

Columns are easily transferable into new architectural settings and are becoming increasingly popular. If you want to use them for structural purposes, check with a structural engineer that they're up to the job; otherwise they make great decorative statements in any room. Stone, wood, cast iron, even plaster – you can find columns made from all sorts of materials. Use them inside the house or out in the garden. Buy four and make them the columns for an outdoor gazebo. Use the base or capital of a stone column as a side table or plant stand. Detailed fragments of a stone column can be mounted on the wall as eye-catching artwork.

EXPERT TOP TIPS

Drew Pritchard: Church furnishings

- Look for the unusual – candleholders, runners, bible stands. Think beyond pews!
- Floors are often very good quality mahogany and oak parquet or tiles.
- Architectural stone is excellent quality and is difficult and expensive to replicate. Even if not whole, stonework always looks decorative and ornate.
- Always try and buy things in complete sets with minimal damage.
- Don't go on your own to demolition sites, as they are exceptionally dangerous; always buy through a dealer.

TRAFALGAR STREET
124 CROMWELL RD. S.W. 7
3 HOUR
F. H. ALLEN
Tailor &

COMMERCIAL, PUBLIC SECTOR AND INDUSTRIAL SALVAGE

Visit any salvage yard and you'll find an abundance of reclaimed material from commercial, public and industrial buildings. Bright red telephone boxes, cinema seats, railway benches, hospital tables, juke boxes, chemist's jars, display cases, farm machinery, school desks, shoe lasts – these are just a tiny fraction of this enormous resource. Many of them are both visually exciting and hugely practical, so how have these architectural treasures ended up on the scrap heap in the first place?

As with any part of the commercial world, there is a constant natural process of updating, refitting and refurbishing. The introduction of different manufacturing methods, novel ways of working and completely new technologies always leaves behind a huge volume of 'redundant' materials. Companies constantly refit their premises in the hope that they appear modern and up to date. New laws and Health and Safety legislation also call for different materials or equipment to be used. And as fashions change, trend-setting shops and restaurants must also refurbish or lose valuable custom.

Take cinemas, for example. Few movie theatres have kept their original interiors and the ones that have are considered a rarity. Cinemas are a cutting-edge, technology-based industry: they want to appear to have the latest state-of-the-art equipment and surroundings. Nostalgia is all well and good for the art-house cinema-goer, but modern audiences demand the latest in sound and picture technology, along with a raft of food counters, entertainments and other conveniences. For the cinema industry it's a case of update every few years or fall behind the times. This constant process of renewal leaves behind a lot of salvage – cinema seats, popcorn makers, old projectors, mirrors and flooring, for example – much of it highly stylized and dramatic, as is often the case with cinema interior design.

Public sector buildings are also regularly refitted. Hospitals, schools, libraries, doctors' surgeries, civic halls and local council buildings receive a lot of wear and tear. Old fixtures and fittings are often just thrown on the skip. Other times, changes in government funding call for the closure or amalgamation of buildings, leaving behind an interesting selection of potential salvage. Some public sector buildings have to modernize to keep up to date. It's no use having antiquated equipment in a state-of-the-art hospital, for example, but that doesn't mean all the porcelain sinks, taps, tables and other hospital salvage need go to waste. Many public sector buildings used very high quality materials in the past, as a reflection of civic pride and the wealth of the country: look out for wood panelling, desks, chairs, doors, fireplaces, marble floors and columns from sources such as these.

As well as this type of salvage, you also find a large amount of industrial salvage – the remnants of a time when Britain was a global manufacturing giant. In the nineteenth century, Britannia really did rule the waves; it controlled much of the world's manufacturing and exported millions of pounds' worth of goods to the farthest corners of the earth. In the twentieth century British industry, unable to keep up with cheaper labour and raw materials from abroad, changed its focus from manufacturing to a service-based economy. Mills, factories and plants closed up and down the country, leaving behind empty buildings and redundant equipment, some of which has ended up in salvage yards. Take the Yorkshire woollen industry, for example. In Victorian times wool-producing cities such as Bradford and Leeds thrived, buoyant on the huge fortunes of mill owners and investors. Unable to keep up with cheap imports, however, the industry all but disappeared in the middle of the twentieth century, leaving behind a profusion of buildings and equipment – wooden looms, signs, huge spools for winding thread, cloth cupboards, chairs, tables, leather buckets, mahogany worktops (from the cloth checking rooms) and enormous baskets (known as skeps) – not to mention all the usual salvageable building materials – wooden floors, hand-made bricks, sanitary ware and roof tiles.

So who buys commercial, public sector and industrial salvage? Fifty years ago it would have been unthinkable for most people to put such items in their homes. No right-minded housewife would have wanted a huge piece of industrial machinery on her carpet or a well-used butcher's block in the kitchen.

The fashion for this type of salvage seems to have grown with the rise in loft living. A few people in the 1960s were avant-garde enough to see the potential

of converted industrial spaces, but it wasn't until the 1980s that the trend really took off. With urban property at a premium, developers and home buyers were looking for cheap alternatives to city housing. Redundant mills and factories offered fantastic potential for a whole new way of living and working, but they also brought with them their own challenges. Most domestic furniture is dwarfed by the huge spaces in loft apartments (the ceilings can be twice as high as normal houses), so creative-minded people turned to large, industrial pieces of salvage for their furniture and fittings. Long factory tables, for example, were and still are highly sought after by salvage hunters. Before the First World War much factory assembly and handwork was done by people sitting at huge, long tables. These were often well made and sturdy, with durable hardwood tops. After years of use, these tables gain a wonderful patina but retain their strength. They're also practical if you want to sit all your friends around a table for an evening meal or barbecue.

The blank canvas of a loft apartment was also the perfect background for the creative use of industrial and commercial salvage – no carpets to dirty, no wallpaper or curtains to clash with. The floors would support heavier pieces of furniture or urban art, while the exposed ceiling supports could hang exciting pieces of industrial machinery. All the conventional rules of domestic interior design were broken. In fact, no self-respecting loft-living 1980s yuppie was without his converted telephone box shower, barber's chair or car-sawn-in-half desk.

SOURCES

There is an infinite variety of reclaimed material from this area: every business, shop, public building and factory has its own specialized equipment and fittings ready to be salvaged. It would be impossible to list all the various types of salvage available, but the categories below give you some clue as to the scope of what's available:

- **Shops.** Every type of shop – chemist, shoemaker, grocer, apothecary, butcher, dressmaker, hairdresser, post office, baker. Jars, bottles, counters, cash registers, window dressing, mangles, hat forms, counters, display cases, dummies, shoe lasts, trade signs, bells, printer's drawers and blocks, lights, drawers, baskets, pallets, clocks, containers, safes, weighing scales, barber's chairs, racks and stands, boxes and advertising.

- **Schools and colleges.** Desks, chairs, blackboards, cupboards, canteen tables, loos and washbasins, gym equipment, lockers, radiators, lecterns, easels, benches, library shelves.
- **Cinemas and theatres.** Seats, mirrors, counters, signs and pictures, doors, projectors, set design, lighting.
- **Factories.** Long factory tables, machinery, clock towers, baskets, chairs, railings, drums and canisters, cogs, shafts, pulleys, anvils.
- **Farms.** Stone troughs, animal feeders, old equipment, barn doors, dairy items, gardening and threshing tools.
- **Circuses and fairs.** Fairground rides, carousel horses, decorative wooden signs, stalls, mirrors.
- **Hospitals.** Sinks, taps, tables, beds, screens, chrome cabinets, trays.
- **Transport.** Railway benches, station clocks, propellers, train seats and tables, bus seats, petrol pumps, garage signs.
- **Shipyards.** Boats, rope, nets, lobster pots, buoys, timber, sails, masts, ship's wheels, compasses, clocks and navigation equipment.
- **Leisure.** Tiles from swimming baths, changing rooms, slot machines, chrome fittings, sports memorabilia, gym equipment, salvage from zoos, museums and exhibitions.
- **Pubs, hotels and restaurants.** Kitchen equipment, ovens, fridges, sinks, neon signs, lettering, pub signs, juke boxes, benches, counters, tables, chairs, towel dispensers, lighting, bars, drinks cabinets, stools, swivel seats, beds, cupboards, sanitary ware.

Most yards will have a selection of the types of salvage above. Industrial and commercial salvage is still relatively cheap, but there are certain areas that have become highly collectable in the last few years. Railwayana, for example, can be very expensive. Locomotive nameplates are the most sought-after item but cast-iron station signs, railway benches and station clocks all fetch healthy prices at auction. The same applies for London Underground salvage.

Another highly collectable area is enamel advertising signs – also known as street jewellery. From the 1880s to the 1950s brightly coloured vitreous enamel signs adorned shops, railway stations and streets up and down the country. They were perfect for advertising: the enamel retained its bright colours far longer than paint and could be cleaned easily. The durability of enamel signs also gave the impression that the manufacturer and his product would last for ever. In the early part of the twentieth century, however, enamel signs fell out of favour, first hit by the Great Depression and then finished off by the banning of steel for advertising during the Second World War. For forty years enamel signs were considered worthless. Nowadays they are back in favour with a vengeance; collectors can't seem to get enough of the nostalgic designs, superb colours and charmingly antiquated language (such as 'Burma Sauce – The Only Sauce I Dare Give Father' or 'OXO – Beef in Brief'), and are prepared to pay hundreds of pounds for the privilege.

Trade signs are also sought after. As early as the seventeenth century, shop owners would advertise their wares with a hanging sign – often carved in wood and painted. Different signs were connected with different trades: the Lamb with hosiers and milliners or the Golden Fleece with mercers, for example. These associations soon became blurred – if a tradesman changed address he would often simply adopt the sign of the old shop and add his own sign – hence the peculiar names that you see across the country: the Leg and Seven Stars, the Eagle and Child, the Shovel and Boot, the Magpie and Crown and so on. Other trade signs had symbolic connotations: traditional pub signs with religious symbols such as the Angel and the Cross Keys reflect the long-standing association of public houses with refreshments for pilgrims. The sheer oddity and historic charm of many antique British trade signs make them really collectable in today's market, especially to American buyers.

As well as individual pieces of commercial salvage, if you are lucky enough you can sometimes find complete shop interiors for sale. Most of these will come up for auction or will be found through a specialist salvage dealer, but it's also a good idea to keep your eye out for old-fashioned shops that look as though they are about to close down or be refurbished. There's no harm in asking the owner whether he'd be prepared to sell you the complete interior. If he will, make sure you supervise the removal of the interior, otherwise you could end up with a very expensive pile of firewood. Look out for old chemist's shops, apothecaries, draper's, grocer's, tobacconist's, printer's and bicycle shops – these often have some of the most interesting display cabinets and chests of drawers, often with lots of little compartments. Shop interiors are also finding their way to Britain from other parts of the world: American diners, French *boucheries* and eastern European bakeries, for example.

Once in a while, auction houses will have a sale completely devoted to commercial salvage. These are an absolute treat. In 1997, for example, Christie's in South Kensington auctioned off the Lost Street Museum, an enormous private collection of antique shop fixtures and contents that had been gathered by two keen collectors, Roger and Pauline de Palma. Lots for sale included the entire contents from a Victorian grocer's shop: display cabinets, enamel signs, mirrors, cash register, counter-top scales, and all manner of antique packaging, tins and advertising. You could also buy entire shop fronts, including a radio shop, tobacconist, lamp and glass shop, toy shop, confectioner, grocer, sales kiosk and a pub as well as three street lamps and numerous vending machines.

USING COMMERCIAL, PUBLIC SECTOR AND INDUSTRIAL SALVAGE

You don't have to have a loft apartment to enjoy commercial, public sector or industrial salvage. In the past decade interior designers have started to incorporate this type of salvage into many smaller-scale projects: butcher's blocks, weighing scales, factory tables and laboratory sinks fit well in modern kitchens, for example, while tailor's dummies, mirrors, shoe lasts and dress racks are eminently reusable remnants from shop refits. Some pieces of commercial and industrial salvage are simply decorative and fun – a six-feet-high hot-dog advertisement or brass lettering from a public house, for example – but most is practical and useful.

In fact, that's one of the greatest bonuses of industrial, commercial and public sector salvage: it was built for a purpose and designed to last. Much of what you'll find in salvage yards is surprisingly well made, frequently using expensive hardwoods or metals. The Victorians, for example, often produced industrial machinery that was not only functional, but also well crafted and highly decorative. They also built things to survive a lot of wear and tear. Think how many bottoms have sat on a nineteenth-century school chair or pew. Or how many times the hands have revolved on an old station clock. Rarely in today's market will you pick up furniture and fittings of such high quality at such reasonable cost.

Much commercial salvage is also highly ornamental. Advertising and shop signs are deliberately designed to catch the eye with their bold colours and fun designs. Life-sized cut-outs of people or objects make whimsical displays, as do shop mannequins and glove-maker's hands. Other commercial salvage represents a snapshot of a particular era – a 1950s motel neon sign or a dramatic Art Deco cinema mirror, for example. By buying commercial salvage you are capturing a piece of social and style history.

Many top designers such as William Morris, Le Corbusier and Ludwig Mies van der Rohe were employed on commercial as well as domestic projects. When any of these important buildings gets demolished (which, thankfully, is happening less and less), at least we can keep some of the pieces of architectural history. Frank Lloyd Wright, for example, designed a fabulous Chicago dance venue called Midway Gardens. Tragically, it was pulled down in the 1920s, but you can still find salvage from it on the open market. For fans of architecture, owning a piece of salvage designed by Wright is as exciting as owning an original painting or sculpture.

Commercial, industrial and public sector salvage can also be really quirky. You could go for something a little bit macabre: a Victorian prison loo or a morgue table, for example. If you're into technology, coloured glass electrical insulators make the most fantastic tea-light holders and drinks glasses. You can even use old massive electrical switchgear to turn on your hi-fi – just make sure you get a qualified electrician to fit it.

Equally, why not indulge the childlike side to your personality? Pick a favourite fairground ride, such as the giant teacups or mini aeroplanes, and plonk it in your garden as an unusual seat. Some salvage is just plain silly – fake lobsters and fruit from food shops, decoy ducks and fairground mirrors, for example – but they can add a real sense of fun to a house.

Outside, why not create garden features from old cider mills or use old hemispherical boilers cut off and up-ended to make summerhouses and sheds? Or, if you love retro design, look out for the wealth of 1950s, 1960s and 1970s salvage on the market: plastic seating, chrome bars, drive-in signs, Formica tables. Create your very own diner or bar at home if you want to go the whole hog. Retro salvage is getting more expensive by the day, so some of these pieces can represent a real investment.

EXPERT TOP TIPS

John Machin, salvage expert: Commercial salvage

- Manufacturing and commercial company auctions often have interesting 'miscellaneous' items going cheaply.
- Buy for yourself – don't follow what's fashionable.
- Purists demand perfection – accept the odd chip or dent and get a better deal.
- Get it free: become a 'skip rat', but ask before you take.
- Office fittings can also look good in the home: for example, storage units and desks.

1033

CROWN & ANCHOR
A
SELECTION OF
WINES & SPIRITS
CELEBRATED ALES
WATNEY COMBE REI

DOORS AND DOOR FURNITURE

When you're wandering around a salvage yard, you'll find two basic types of doors. Battened doors are the simple, rustic-looking doors made from vertical planks of timber nailed together and strengthened with horizontal pieces of wood or 'ledges'. This type of door has been in widespread use since the Middle Ages and is still a common feature in country cottages, rural buildings and outhouses. Battened doors also enjoyed a revival through the Arts and Crafts movement – their rudimentary style and rural character appealed to the movement's love affair with the past.

A battened door will usually be held in place with simple wrought strap hinges (long external hinges that stretch across the door) and have a simple ring handle. Very early battened doors were made of thick hardwood such as oak and constructed with hand-sawn lengths of timber. This often means that they have vertical planks of different widths. As sawing techniques improved and became more standardized, battened door planks became more uniform in width. Battened doors are also known as 'ledged' doors. If they have extra diagonal wooden braces on the back, as appeared in the nineteenth century, they are known as 'ledged and braced' doors.

The other type of door is the panelled door. Familiar to anyone who owns a Victorian or Georgian house, this type of door has actually been in existence for hundreds of years but only became common at the beginning of the eighteenth century. Panel doors overtook battened doors in popularity for a number of reasons. Before the 1700s panel doors were too expensive for the majority of home owners. Panel doors require more sophisticated carpentry skills than battened doors, namely mortise and tenon joints, and this made them prohibitively expensive to anyone but the wealthy. However, with the introduction of semi-automated wood-cutting technology and quicker assembly in the 1700s, panel doors could be made more rapidly and cheaply than before. Panel doors also require less wood than battened doors – by the end of the 1700s British hardwoods were becoming rarer and more expensive, so the panel door became a useful, more practical alternative, especially if it was constructed from abundant softwoods such as pine.

Fashions in panelled door design have fluctuated, but the basic construction has stayed the same: panels of wood secured loosely in a mortise and tenon frame. The number of panels has varied over the years from two large panels in a typical late seventeenth-century door to six in a Georgian door and four in a Victorian. The use of stained and coloured glass became popular in Victorian times as a way to bring light into the dark, imposing interiors of the day, without losing any privacy. The Art Nouveau movement copied and improved on this tradition, producing beautiful, fluid examples.

DOOR FURNITURE

Both cast iron and wrought iron have been around for thousands of years. Wrought iron is worked in its solid state, heated and hammered into shape. Cast iron, on the other hand, is liquid and poured into moulds or casts to set. Wrought iron was used predominantly for door furniture until the Georgian period, when the introduction of new blast furnaces helped make cast iron a cheaper alternative to hand-made wrought iron. Only in the nineteenth century did brass become the metal of choice for most door furniture, alongside porcelain and glazed earthenware. In an interesting reversal, however, from the 1900s onwards, iron once again became popular due, in part, to a shortage of domestic staff to keep brass polished.

LOCKS AND FINGERPLATES Early buildings were kept locked with a heavy timber bar jammed behind the battened door. Wooden or iron bolts were used in grander houses, but it wasn't until the seventeenth century that you start to see proper surface-mounted door locks. By the early nineteenth century these had developed into the more sophisticated mortise locks (set inside the frame of the door). Fingerplates were also introduced in the seventeenth century to protect painted doors from grubby fingermarks and carried on being popular right up until the 1940s.

HINGES Before the 1700s most hinges were strap hinges, mounted on the surface of the door, clearly visible and part of the door's design. With the rising popularity of panel doors in the eighteenth century, hinges became similar to the design we have today – smaller in size, cut into the side of the door and hidden from view.

LETTERBOXES If you've bought a front door that predates the middle of the nineteenth century it probably won't have a letterbox, unless one has been added at a later date. Letterboxes only became common after the introduction of pre-paid postage stamps and regular door-to-door deliveries in the 1840s.

KNOCKERS Door knockers were well established by the beginning of the 1700s and can add real character to a door. Look out for animal heads or feet or other unusual designs, as well as the simpler, more subtle examples.

SOURCES

A dealer specializing in salvaged doors will have hundreds if not thousands of doors to choose from, covering most periods and styles. The doors should be stacked upright, like books on a library shelf, so you can leaf through at your leisure. Ideally the doors should be grouped in some kind of order (age, price or style). Even though they may also have the sizes written on them, always re-measure for yourself to make sure before you buy.

Unless you are confident about what's underneath, avoid painted doors. The paint may be hiding a damaged surface or split panels, or, even worse, a panel that has been replaced with a cheap new material. Varnished doors, on the other hand, are always popular. They can be easily stripped, sanded and revarnished, but you can see exactly what you're getting when you buy. Most doors in salvage yards, however, will have already been stripped for you. All you need to do is decide on the finish.

Look for a door with solid construction, free from damp or evidence of wood rot. Check for intact mouldings and beading, and, even better, the original door furniture. If the door hasn't got its furniture, however, don't panic; the salvage yard will probably have a good selection of antique hardware that will match the style and era of your door. Look out for knobs, lock plates, pull rings, letterboxes, numbers – anything that will add period, authentic character to your door.

Be careful if you plan on buying a panelled door that was 'converted' in the 1960s. Although, in theory, the thin layer of hardboard should have protected the door underneath, as often as not the mouldings and any other protruding decoration will have been ripped off and you'll end up disappointed.

RESTORATION AND CLEANING

For the past twenty years it has been common practice to send your softwood Victorian doors away to be dipped. And, while stripped wood can look really beautiful, it's worth noting the following things:

- Victorian pine doors were never made to be bare wood – softwood was considered too bland to be left unfinished.
- The process of dipping a door in a bath of sodium hydroxide can potentially raise the grain and cause the joints to weaken by dissolving the glue that helps hold them together.
- Dipping a door thoroughly soaks it; it can take months for a dipped door to dry out completely and, if you try to rush the process, it can cause the door to warp and split.

If you want to be historically accurate in a Victorian home, the doors should be painted. Softwood doors on the ground floor were often grained to simulate oak or mahogany, while all the upstairs doors would be painted plain colours. Choose matt or semi-gloss paint; high gloss is a recent invention. If you want to be really authentic, the panels on many Victorian internal doors were decorated with stencils or paper.

If you do want stripped doors, however, make sure that you take them to a reputable company who will dip them and then remove the residue with pressure washing. It's also important that the doors are left to dry very slowly and, if possible, placed into dehumidified storage until the drying process is finished. Alternatively, you can manually strip off the paint using a paint stripper. Avoid using a heat gun as this can burn the wood.

N.B. If you are hand stripping an old door be aware that most pre-1950s paint contained lead. Lead is very dangerous whether ingested or inhaled so you have to take great care if you want to remove it, especially around children and pregnant women. Use methods that don't create dust or fumes and remember that solvent-free, water-based paint removers are now available – ask your DIY store for details. For more information on safely removing old lead paint, visit the relevant section on the Department of the Environment's website, www.doeni.gov.uk.

Be careful not to buy a warped door. Look down the width and length to make sure the line of the door is straight. If you do buy a warped door by mistake, or find yourself given a warped door that you'd like to use, SPAB (the Society for the Protection of Ancient Buildings) have the following advice:

- Counteract the warp in a door by fitting an extra hinge, repositioning an existing hinge or using a simple barrel bolt to pull the distortion into line. Alternatively, the stop on the frame can sometimes be carefully trimmed or relocated.
- If that doesn't work, a door can sometimes be laid flat and straightened by applying an even load for several days or weeks.

Wood is a living thing and continues to expand and contract its entire life. In panelled doors the inner panels are deliberately loose within the frame to allow for this movement. If, for some reason, the panels become stuck or have too many layers of paints on them they can split. If you can manage to re-loosen the split panel, the two halves can just be re-glued. If the panel can't be re-loosened, thin wood slices will have to be glued into the split and sanded flush with the face of the panel.

If you have a hardwood door, don't be tempted to paint it. Bring out its natural grain and beauty with an appropriate wax or oil finish.

For more information on restoring and maintaining Victorian and Edwardian doors, the Victorian Society (www.victorian-society.org.uk) publishes a handy guide containing sections on history and styles, reinstating lost doors, door furniture and security.

USING DOOR SALVAGE

If you want to use a salvaged door it's easier to buy a door first and tailor the doorway to match. If you plan to add salvaged doors to an extension, for example, pick your reclaimed doors before you embark on the build and get your architect to design door openings to fit.

In existing houses this can make life a little tricky, as you don't want to alter all the door openings to fit your new reclaimed doors. As a general rule, look

for a door that matches or is only slightly larger than the door opening. You can't cut a panelled door down by very much – you'll weaken the structure and ruin the proportions. If you've found a panelled door that you absolutely love, however, and you're desperate to put it in an existing doorway that is too small or too big, consult a carpenter or builder about altering the frame or size of the doorway.

Battened doors are more easily adapted than panelled doors, especially if it's just a matter of trimming a centimetre or two off the top or bottom (taking care to keep the proportions right). It's more difficult if you want to narrow the door, as trimming it lengthways will leave you with odd-looking planks of different thicknesses.

With both panelled and battened doors the best advice is to measure the door opening very carefully before you go shopping and get the salvage yard owner to help you choose suitable doors – they will know which ones can be slightly altered and which will have to remain totally unchanged. If you're getting rid of all the doors in your house consider asking the dealer if he'll take your old doors in part exchange. Early to mid twentieth-century doors are becoming collectable, while 1960s and 1970s retro doors have kitsch appeal for the younger market.

If you're using salvaged doors in a restoration project, it's vital that you choose the appropriate style. Victorian panel doors will look out of place in an early farmhouse, as will battened doors in the downstairs of a Georgian villa. Equally, make sure you use internal and external doors in their correct positions: internal doors are too flimsy to be used as external doors and vice versa. And don't forget the architrave – use the appropriate moulding for the door surround: a battened door should have a very simple, plain door opening whereas a Victorian or Georgian door will need decorative moulding. Make sure the door furniture matches; battened doors shouldn't really have brass fittings, for example.

While there's no end of vintage doors to be found in salvage yards, it's not always easy to find more than four or five in the same style (groups of doors are called suites). Rather than spend months trying to find a whole houseful of identical doors, why not make the most of the situation and choose deliberately different doors? Even if they need to be from the same era, it won't look odd to have slightly differently decorated doors. If anything, it will add to the individuality and character of your home.

As well as the obvious one, reclaimed doors can also be put to good use in other ways around the home. Turned sideways, panelled doors are an inexpensive way to create handsome wall panelling. Just make sure that you choose doors with panels that are the same width.

Doors laid on their sides also make fantastic headboards and footboards for a bed, and a door split lengthways will make the side panels. A battened door can easily be converted into a country-style kitchen table – just find yourself four carved wooden legs.

Tall thin doors, such as French or cupboard doors, look fantastic hinged together in threes or fives to make a decorative screen. Use glazed doors and the screen becomes a useful room divider that won't block out too much light.

A door split lengthways will make the back and seat of a garden bench or futon. Another door can make the sides and armrests. Smaller panelled cabinet doors make great blanket boxes or toy boxes or, with the centre panel removed, can be turned into chunky picture frames and mirror surrounds. Even door furniture has other uses: door knobs attached to a plank of wood make an instant coat rack, for example. Wooden, porcelain or brass door knobs also make cheap but decorative curtain finials.

EXPERT TOP TIPS

Amanda Spencer of Period Pine Doors: Doors

- Always buy stripped doors. Painted doors can hide a multitude of sins.
- Don't get rid of all the bumps and knots – flaws add to the character of a door.
- If you want an unusual sized door, remember that antique doors come in all shapes, thicknesses and sizes.
- If you need doors for a new build or extension, it's easier to buy the doors first and build the door casings around them.
- Ask your salvage dealer if they offer sanding, stripping, filling and waxing – it will save you lots of time and energy.

FIREPLACES AND RADIATORS

Early medieval buildings didn't have fireplaces as such. Whether you lived in a manor house or a modest cottage, an open hearth in the middle of the room would be used for both heating and cooking. Smoke simply rose up to the ceiling and left the building via a hole in the roof, taking dust and bad smells with it. The ashes from the fire were allowed to build up on the floor and to start a new fire you'd pile up logs like a small bonfire or, in wealthier homes, use wrought andirons to support the logs while they burned. These metal andirons would also serve as the basis for a roasting spit.

Despite being in existence for hundreds of years, this type of fire had its obvious problems. It wasn't always easy to keep the flames under control – which was no good if you lived in a tinder-dry timber-framed home – and smoke often filled the room, making it difficult to breathe or see.

The twelfth and thirteenth centuries saw the introduction of simple wattle and daub canopies or hoods above the hearth, to direct the smoke upwards. Rather than try to suspend a canopy in the middle of the room, it was easier to attach the canopy to an end wall and have the fire underneath. The basic wall hearth had arrived.

From the 1500s onwards, methods of construction and flue design improved and the open fire and wattle and daub canopy was slowly replaced by a more sophisticated system. The fire became recessed into a rectangular hole in the wall and the smoke funnelled away by a purpose-built brick or stone chimney stack. The opening was spanned by an oak beam or stone lintel and there would often be an area to sit by the fire and keep warm (also known as an inglenook fireplace).

Although initially a purely functional and utilitarian object, the hearth soon became a focus for ornamentation and design flair – a chance for home owners to express their status and fashion sense. The mantelpiece, which most fireplaces have nowadays, came out of the Renaissance period. Taking their inspiration from Roman columns and pilasters, and other elements of classical design, Renaissance designers added mantelpieces – stone, wood or marble frames – around the opening of the fire. The area above the fireplace, the overmantel, also became a focus for elaborate decoration and a chance to show off one's wealth; the fancier the carving, the fancier the owner.

By the end of the 1700s the nation's taste moved towards less ornate fireplaces – still classical in design and influence, but without so many of the extravagances of previous years. Fireplace recesses were usually square with simple moulding in wood, stone, marble or painted plaster. These subtle, lighter Adam-style structures became the model for future mantelpieces and still remain one of the most popular reproduction designs today.

One of the major changes at this time was the introduction of coal as a freely available fuel source. Hotter and more efficient than wooden logs, coal needs different combustion conditions to burn effectively. Large, wide fireplaces gave way to smaller, narrower hearths, and the andirons were rejected in favour of the iron fire grate or coal basket. To prevent hot lumps of coal or cinders falling out of the basket and rolling into the room, metal and wood fenders were also introduced and soon became a design statement in themselves.

In the late eighteenth and early nineteenth centuries, as the Industrial Revolution swept across Britain, changes in mass production altered the design and technology of the mantelpiece. Marble chimneypieces could be pre-fabricated and pieced together, making them cheaper to buy and quicker to produce. Cast-iron metal fire-surrounds also became common; they were inexpensive to manufacture and easily moulded into the latest revivalist fashion. The Victorians were passionate about the past and embraced every historical era of design, from neo-classical to Gothic revival and back again.

A Victorian middle-class home could now splash out on a marble, stone or hardwood mantelpiece for the living room – the public 'showroom' of the house – and use cheaper cast-iron or painted softwood mantelpieces in the less important private rooms such as bedrooms and bathrooms. In fact, as methods of cast iron manufacture improved throughout the nineteenth century and decorations became more intricate and detailed, even the wealthier households embraced the metal mantelpiece as an eminently suitable alternative to marble and hardwood. Plate glass also became widely available during this time, and

we see the introduction of the overmantel mirror, which gradually eclipsed the carved wooden or stone overmantel.

The Victorians were gloriously compulsive collectors. Every room in a house was filled to the brim with ornaments, clocks, candles and picture frames, not to mention furniture, stuffed animals, plants, cushions and fabrics. The fireplace became an ideal place to display even more knick-knacks and necessitated the widening of the mantel (the shelf part of a mantelpiece) to incorporate as much clutter as possible.

Delft tiles had been used to decorate the hearth area as early as the seventeenth century, but it was the Victorians who really went to town with the idea. Mass-produced tiles in an endless range of patterns, pictures and motifs could sit neatly in the fireplace slip and add another source of busy ornamentation to the room scheme. A few years later, the Arts and Crafts movement of the latter half of the nineteenth century despaired at the mass-produced goods and often poor workmanship of the Victorian age, and looked to medieval times for inspiration, patterns and techniques. It's at this time that we see the reintroduction of medieval-style fireplace hoods and inglenooks, and the use of individual hand-painted fire slip tiles.

The Art Nouveau period that straddled the end of the 1800s and the beginning of the 1900s wanted to create a fresh new style, one that had no design connections to the past. Cast-iron fireplaces were given the Art Nouveau make-over: sweeping long curves of ornamentation, sensual naturalistic designs and slip tiles with organically inspired motifs.

After the Second World War the fireplace suffered a double whammy. Central heating, which up until then had been the preserve of the wealthier classes and big institutions, became cheap enough for many people to enjoy. Electric fires – clean, easy to use and instant – also became popular. People no longer had to lug coal and wood around the house and clean out sooty, dirty fireplaces. Stripped of its function, the fireplace became obsolete, a sign of old technology and outdated ideas. Many were ripped out of old houses and chucked on the rubbish heap, other fireplaces simply boarded over and forgotten about.

During the 1970s, however, the fireplace once again became a feature of the home. The introduction of the 'living flame' gas fire necessitated a return for fire-surrounds, while many people realized that although central heating was providing a reliable background temperature, it wasn't providing a focal point. Nowadays, a house without at least one fireplace is considered unusual. Property developers will tell you that a hearth will score big points with any potential buyer and, if missing, reinstating the original fireplaces can add significant value to a property. Even more interesting, however, has been the resurgence of interest in real fires and wood-burning stoves. People are deliberately going back to simple, age-old technology. A fake fire will provide all the heat you need. But nothing compares to the comforting, companionable sound and intrinsic appeal of a roaring, crackling open log fire.

SOURCES

FIREPLACES

The fireplace has to be the most popular piece of architectural salvage and a really easy way to put history back into your house. Many people who are now reclamation addicts will tell you their first purchase was a mantelpiece – easy to find, easy to clean and easy to fit.

Most salvage centres have a stack of fireplaces and many yards specialize in locating and restoring them. If you have a very definite idea about what kind of fireplace you want, or need technical advice, a specialist dealer should be your first port of call. If your needs are fairly straightforward, however, you should be able to pick up a real bargain at most salvage yards.

The durability of fireplaces is one of their most attractive features and it's not difficult to pick up a mantelpiece that's one, two or even three hundred years old. You'll find a huge range of materials – marble, stone, wood, cast iron – in a wide variety of finishes: burnished, painted, marbled or plain. Look for fireplace accessories, too. These can add real character to your room. Keep an eye out for interesting decorative tiles, especially Delft or hand-painted Victorian examples. Don't worry about getting a matching set if you're strapped for cash – a patchwork effect of antique tiles can look spectacular. And don't forget wrought andirons, metal coal baskets, bellows, tongs, shovels, fire screens and spark guards, fenders, brushes and overmantel mirrors.

Reproduction mantelpieces are also commonplace. Most are good quality and represent an easy way to get the vintage look. Others have been known to twist, crack, split and even melt when in use. On cheaper wood reproductions, for example, the decorations can be made from a different wood than the body of the fire-surround. As soon as the fireplace is subjected to heat, the two woods dry at a different rate and the decorations can fall off. The key is to know what you're getting. A repro is fine – unless it has been sold to you as an original. It's not always easy to tell them apart, but there are a few things to look for:

- **Check the back of the fireplace.** If the screws are new it's a tell-tale sign that the fireplace is new or has had some piece replaced.
- **Usually there will be a name on a cast-iron fireplace.** On old fireplaces the name will be indented. On reproductions the mark will stand proud.
- **Check the detail closely.** On an original cast-iron fireplace the detail should be sharp and clear. Reproductions are made from moulds taken from the original and some of the detail is always lost in the process.

RADIATORS

Chunky Victorian cast-iron radiators can often be picked up for very reasonable prices at a salvage yard and look much more handsome than their white, rather anaemic modern counterparts. They mostly come from school buildings, churches or grander homes, as central heating didn't become widespread until well into the twentieth century.

You have a choice with reclaimed radiators: buy them unrestored and do any repairs yourself or get a ready-to-radiate, fully refurbished reconditioned model. The cost difference is significant, but unless you have access to specialist help and equipment, you may end up paying as much to restore it yourself as buying a fully reconditioned radiator.

This is why. About one in ten unrestored radiators don't work properly. It can be almost impossible to tell from the outside what the condition of the radiator is on the inside. Old radiators may have small cracks or leaks that you can't see at first glance. Radiator gaskets, which seal the joints between the sections, are particularly susceptible to damage so, to ensure it is serviceable, a salvaged radiator should be flushed out with a pump system and pressure-tested. The connections between modern pipework and old cast-iron radiators may be different sizes, so it's also important that you make sure a plumber can source the right parts to make it compatible with a modern central heating system. At the very least, ask the dealer whether your radiators have been pressure-tested.

On top of this, your salvaged radiator might be covered in layers of old paint. Not only can this spoil its appearance but it can make it less efficient at radiating heat. To have paint removed, your radiator will need sandblasting and repainting. This is best done by an expert.

RESTORATION AND CLEANING

FIREPLACES

The process of cleaning and restoring a fireplace depends on the material.

CAST IRON To remove dirt from a cast-iron fireplace, use methylated spirit on a soft cloth (don't use water – it can encourage rust to form). Polish the grate to a dull sheen using a small amount of appropriate polish (such as Liberon Iron Paste or Zebo polish) and a brush. Buff with a soft cloth after two hours. If it needs more than just a clean, only use high-temperature stove paint.

If your cast-iron fireplace has been painted and you can remove the paint, use a paint stripper: Nitromors is a well-known brand and Homestrip from Eco Solutions (www.ecosolutions.co.uk) is a non-toxic, non-flammable, solvent-free option. Apply the stripper as instructed by the manufacturer. Scrape off the melted paint with a spatula and work on fine detail with a nylon-bristle brush.

Incomplete cast-iron fireplaces can be problematic, as it's not always easy to find spare parts or economical to get them repaired.

SLATE For slate, use a soft brush and a solution of gentle washing-up liquid and warm water. Wipe off any excess liquid immediately and use a dry cloth to finish. If you want to seal the slate, to prevent fingermarks and other stains, you can use linseed oil, but not in areas that get hot.

MARBLE To clean marble, use a mild pH-neutral cleanser specifically developed for stone surfaces. High alkaline or mild acidic cleaners can dull, stain or even damage the finish. Always do a patch test first. You should be careful when cleaning a marble fireplace: sometimes slate or stone fireplaces were skilfully painted to resemble marble and if you over-clean you could find yourself with a shock!

Marble staining will need poulticing. A paste, such as Sepiolite, is prepared using the appropriate poultice, applied to the stain and left to dry. As the poultice dries it draws out the stain. If poulticing doesn't work, you can consult a professional to 'hone' or grind the surface down to a fresh layer of stone and then re-polish. Fine cracks and spalling (chips or fragments in the face of the stone) can be repaired with a suitable filler, usually a polyester, epoxy or cement-based filler coloured to match the stone.

WOOD To clean a fine wood fire-surround, mix four parts white spirit with one part linseed oil. Apply with a cloth and wipe dry. Apply beeswax to finish. (N.B. Both white spirit and linseed oil are highly flammable – do not apply these substances while the fireplace is operational.)

If you're planning to strip your wood fireplace, make sure that any carved decoration is wood and not gesso (plaster of Paris); the gesso may not survive the dipping process and you could be left with a ruined fireplace.

If your fireplace has any significant damage you'll need to get a specialist restorer on the job. Major repairs to marble, such as taking out a whole section and replacing it, will need the services of an expert stonemason, for example, or cast-iron fireplaces that need welding or shot-blasting will have to be mended in a workshop.

RADIATORS

If you're confident that your salvaged radiators are in good working order, they might just need a lick of paint to get them looking as good as new. To paint an old-fashioned radiator you'll need a brush that tapers to a point (to get into all the inner recesses). Hamilton Acorn has a range of brushes that taper to a point (available at B&Q). Working from the least accessible parts of the radiator outwards, prepare the surface with a fine sandpaper. Wipe down with a damp cloth. Undercoat and then apply two thin coats of hardwearing radiator paint, which can withstand changes in temperature.

USING FIREPLACE AND RADIATOR SALVAGE

Fireplaces are one of the few pieces of architectural salvage that have no other function than their intended one. In other words, you can't do anything else with a mantelpiece apart from put it around a hearth. That doesn't mean, however, that you can't be creative with your choice and use of fire-surround.

The variety of fireplaces on offer is astonishing – from the primitive inglenook to the hugely ornate high Victorian. You won't fail to find something to suit your taste and house. If you're restoring a home or replacing missing fireplaces, take care to choose the right fire-surround, grate and tiles for the period. If you're renovating a listed building, you might have to choose a very specific type of fireplace to fit in with the rest of the house. In fact, even if you want to reinstate the fireplace as it was originally built, you may need to apply for listed building consent. Call the conservation officer in your local planning department for advice. Consider the proportion of the room against the fireplace. And don't mix up the various elements: a 1930s fire-surround will look odd around a Georgian hearth. The idea is to make it appear as if the fireplace has always been there, rather than added at a later date. A fireplace is the focal point of any room and will get a lot of attention: make sure that it enhances the room rather than sticks out like a sore thumb.

You'll also need to check that your fireplace and chimney are practical and suitable for the job in hand. One of the most common problems is that the fireplace isn't compatible with the chimney and causes smoking. Check with a builder. You'll probably need to get someone to sweep the chimney, and it may even need lining.

If, however, you are a little more flexible with your design scheme, why not think about doing something unusual with your fireplace? Modern loft spaces with their austere brick walls and industrial design, for example, can provide a stunning background for a huge over-the-top marble neo-classical fire-surround. Play around with scale and proportion. For the ultimate in decadence and comfort, put a working fire in your bedroom or bathroom – there's nothing better than taking a relaxing warm bath watching real flames flicker in a fireplace.

EXPERT TOP TIPS

Anthony Reeve of LASSCO: Fireplaces

- The size of the room will determine the size of the fireplace – trust your eye on this.
- The width of the chimneybreast is the measurement you really need – not necessarily the size of the hole, which can often be altered.
- Find out if neighbouring houses in the same street still have their period fireplaces as a guide.
- If you want to know how your chosen fireplace will look scale-wise, make a mock-up using newspaper or cardboard or draw it on the wall.
- Look for the chimneypiece first and then find a grate to fit it.

NO DOGS
ALLOWED

FLOORS AND ROOFS

FLOORS

From beaten earth and bricks, to flagstones, slate, marble, fired tiles and pebbles, you can find an infinite variety of solid flooring, and much of it has been in use since early times. Beaten-earth floors were common in most homes up until the seventeenth century, but a few artisans' cottages had earth, clay and oxblood mixed together to make hard floors not dissimilar in appearance to quarry tiles.

By the seventeenth and eighteenth centuries stone flags, quarry tiles and bricks became popular floor coverings, but there still existed a huge amount of regional variation, access to materials being restricted by high transport costs and proximity to quarries or brickworks. Smarter houses would often choose more expensive materials, such as polished limestone and imported marble, both of which didn't become more affordable until the nineteenth century and even then remained prestige items.

Before the eighteenth century, timber wasn't often used as a covering for ground floors since it was too vulnerable to rot. Timber did, however, make a useful floor covering for the upper rooms. Early floorboards were wonderfully wide and chunky, often made from tough, beautiful woods such as oak and elm. Each floorboard could be more than 30cm in width and, because they were hand-sawn, irregular in appearance. In fact, early floorboards were often so substantial that they could serve as the floor, ceiling and floor joists all in one.

After the eighteenth century, with improvements in building techniques that dealt with ground-floor damp, timber became a viable option for downstairs as well as upstairs. Unfortunately, by this time supplies of native oak and elm had already become scarce and too expensive for most pockets, hence the introduction of softwoods such as pine and red fir as affordable alternatives.

As machine wood sawing came into use floorboards also became standard widths and much narrower – nearer 10cm.

In the late seventeenth and early eighteenth centuries there was a trend in well-to-do houses to use parquet flooring (wooden blocks laid in geometric patterns), but it wasn't until the Victorian era that the fashion for parquetry and marquetry (inlaying different colours of wood to create a pattern or picture) really took off. It was also the Victorians who made encaustic floor tiles popular, despite the fact that earthenware floor tiles had been around since the Middle Ages. In fact, you can rip up the carpet in almost any Victorian hallway or porch and chances are you'll find a highly patterned and colourful tiled floor waiting to be revealed. The word 'encaustic' comes from the Greek *enkaustikos*, meaning 'burnt in', and reflects the fact that the bright decorations are bonded to the tile by heat.

ROOFS

Although we traditionally think of thatch as the oldest roofing material, both fired clay and slate tiles have been used since Roman times. Because clay and slate are natural materials, their colour and composition changes depending on where they are found. The clay of Cambridgeshire produces warm yellowy-brown tiles, while in Sussex, for example, you find redder tiles. The wood-firing process also gave clay roof tiles an individuality: the nearer they were to the fire, the deeper the colour. This kind of colour variation is part of what gives our old buildings so much character.

It's the same with slate. Welsh slates can have notable differences in colour, even if they come from the same quarry. The slates used as roofing material in the Lake District have wonderful surface variations and texture.

Throughout history, different parts of the country have used different roofing materials to protect and enhance their homes. Each region has its own strong vernacular tradition, so the first thing you'll need to do, before you buy any reclaimed roof tiles, is to do your research. Find the right materials for your area and the age of your building. Concrete roof tiles, for example, didn't really appear in Britain until the end of the First World War. Equally, modern clay tiles are fired in a gas kiln, not a wood kiln, and this gives them a very uniform appearance that can look out of character on a period property. Look at the other period houses in your area and consult a roofer who understands conservation issues.

SOURCES

FLOORS

Second-hand floorboards, especially if they're made of oak or elm and have a nice patina, will enhance any home and make a durable, practical floor surface. It's worth going to a specialist dealer if you want a large quantity of floorboards, as they will have a bigger selection to choose from. The dealer should also have flooring bundled together from the same source, so you'll get an even matching floor rather than a jumble of planks.

Reclaimed floorboards will come from a variety of sources: grand houses often yield beautiful and expensive hardwood examples. But you can also pick up floorboards from old schools, factories and other places, which represent great value for money but still have a wonderful patina. One thing worth bearing in mind, however, is that reclaimed floorboards from woollen mills may be impregnated with a strong lanolin smell – so make sure you have a good sniff of your floorboards before you buy!

If tar was used to lay the floorboards originally, put your salvaged boards in a cold store and this should cause the tar to drop off. Also, never store floorboards 'face to back' as the nails or tar on the back of one board may damage the face of the adjacent board. Equally, make sure the mastic material that was used originally to lay a parquet floor can be satisfactorily removed.

Above all, make sure your floorboards are dry when you buy them and, if possible, lay reclaimed flooring loose in the room for a few months before fixing them down. This will allow the wood to adapt to the surrounding temperature and moisture levels. There's nothing worse than carefully laying a wooden floor, only to watch it warp over the following weeks.

ROOFS

Pre-1940s clay tiles can be pricey, but the effect is marvellous. They're also good quality and likely to last substantially longer than poorer quality post-war clay tiles. If you're looking to match up existing tiles, take a sample with you – it's the only way you'll get an accurate match both for size and colour. Don't be afraid to pick up and examine the tiles: look at them closely and check for any cracks, chips or splits.

Old clay tiles are usually fairly durable and you often only need to replace a few broken tiles rather than entirely re-roof the property. Check that the tiles are fixed in the same way, e.g. whether your tiles are peg tiles or nib tiles. Expect to pay double or triple what you'd pay for modern clay tiles – it might seem a lot, but the aesthetic and qualitative difference is huge. When you're buying a bulk pallet of tiles, always check the second row because all the good ones tend to be put on the top.

If possible, use the same roofer to remove the tiles and put the new ones on. One extra thought: cowboy roofers have been known to remove perfectly good expensive roof tiles and replace them with cheap modern tiles. They then sell the old tiles and make a tidy profit at your expense. If you own a period property and your roofer tells you that he needs to replace the roof, ask to go up the scaffolding to see for yourself. And go up at the end of the job to inspect the finished result.

With slate tiles, any salvaged tiles should also match the type, colour and texture of your own roof. Many of the old slate quarries have closed, so reclaimed tiles are often a good option. When you're re-covering your slate roof, you should also try to reuse as many existing tiles as possible and blend the existing and salvaged ones together.

RESTORATION AND CLEANING

FLOORS

Using salvaged timber floors doesn't really represent any major challenges – the hardest part is matching the old and new wood if you are patching an existing floor. Remember that softwoods darken with exposure to sunlight.

If you want to buy an entire floor, before you part with your money decide where in the house it will go as this affects the choice of material. Areas of heavy wear and tear, such as hallways and landings, are better suited to the more durable hardwoods like elm and oak, whereas softwoods will suffice in low-wear rooms such as bedrooms and attics.

If you've bought an expensive complete antique floor, it will be easier if each plank has been marked and numbered so you can lay the floor in exactly the same position back home. This is because old floors will have different areas of wear and colour which need to be re-created on installation. Take care if you're buying a floor *in situ* – it might be difficult to tell whether the underside of the

floorboards are affected by dry or wet rot. Dry rot has a distinctive damp, musty smell; if you think a floor has dry rot, don't buy it under any circumstances as once installed it will spread to other parts of the house, causing untreatable and expensive damage to other timber, plaster and brickwork.

Once you've installed your new wood floor, you'll need to keep it shipshape. Rather than polish it, which can leave a wooden floor perilously shiny, the common practice in the eighteenth century was dry-scrubbing. It's rather laborious: you first dry-scrub the floor with damp sand. After sweeping up the sand, you then brush in dried herbs and petals and sweep them up also. You're left with a sweet-smelling and not too shiny floor.

However, if you haven't got time for sand and sweet herbs (and let's face it, who has?), you can maintain a wood floor with little more than regular sweeping with a soft bristled broom or vacuuming with a soft floor attachment. You also should clean your floors periodically with a professional wood floor cleaning product. Don't wet-mop a wood floor – it can damage the wood – and try not to overwax the floor, as this can lead to a dull finish.

To repair any stains or scratches on a wood floor, try the following: if your floor has a wax finish, you can remove any dark spots or stains by gently rubbing with very fine steel wool and then re-waxing the affected area. If the spot persists, apply a small amount of vinegar and allow it to soak for an hour. Rinse with a damp cloth, wipe dry and smooth with fine sandpaper. Stain and re-wax. If it's a water stain, rub the spot with very fine wire wool, stain and then re-wax. For greasy or oily stains, use kitchen soap with a high lye content and then re-buff. Finally, if you have too much wax (called 'wax build-up'), you can strip away the old wax with a specially designed floor product and very fine steel wool. If the floor is varnished, choose a touch-up kit made specifically for urethane finishes. If the stain is in the wood, not the varnish, you may have to sand it away and re-varnish, but be careful that this doesn't take away the character of the floorboards.

For grease-stained quarry tiles and stone floors, use a solution of 1 tablespoon caustic soda in a 4.5 litre bucket of water. Rinse well afterwards.

You'll need a specially designed cleaner on marble floors, but slate will respond well to a home-made cleaner of one part linseed oil to four parts white spirit. You can hire a polishing machine for most smooth stone floors such as marble and limestone, but take care not to over-polish and end up with a skating rink for a floor.

Encaustic tiles will come clean with a gentle detergent and some warm water. You'll need to seal them with a specialist sealant (ask at your local tile retailer) to prevent any further damage.

ROOF TILES

There's little to be done about restoring or cleaning roof tiles. Either they are intact or they're not. If you can see that a clay or slate tile has split into thin layers (a process known as delamination) the tile can't be repaired and should be discarded. Applying waterproof paint or any other modern coating is also a no-no – if a roof tile is sound, it shouldn't need any extra treatment.

EXPERT TOP TIPS

John Rawlinson of Original Architectural Antiques: Roofs

- Make sure roof tiles match existing tiles.
- Make sure tiles do not have frost damage.
- Check that when using timber it will be structurally sound enough to take the weight of the tiles and the correct size.
- Make sure that the beams suit the property.
- Timber should be treated before use.

GARDEN SALVAGE

If you visit the House of Vettii in Pompeii, you will see the perfect re-creation of a Roman garden built on the very site where it was destroyed by the famous volcanic eruption of AD 79. From excavations like this and others such as Fishbourne Palace in Sussex, archaeologists have learned that the Romans were wonderful garden designers, keen to use statues, fountains and other ornaments to bring a sense of structure and surprise to a garden. Their ornate and formal gardens, planted with low box hedges, would have been full of marble and bronze statues, fountains that spouted water, garden seats and urns on pedestals – all designed to create focal points and dazzling all-year-round interest.

A thousand years later, the western world began once again to take an interest in classical garden design. Inspired by the ancient ruins that surrounded many of their great cities, Italian garden designers of the Renaissance revived the Roman fashion for elaborate stone statues and other garden ornaments, set within a formal garden layout. The trend caught on and soon many of the grand houses of Europe had strong elements of classical design and ornament in their grounds. In England, for instance, the Tudors followed the Italian example and, for the first time since the Romans left, sundials and statues were the must-have garden ornaments.

The boom in classical sculpture and ornament, however, came during the eighteenth and nineteenth centuries. Young members of aristocratic families were expected to finish their education with a Grand Tour of Europe, a trip that often included some of the most important ruins and archaeological sites: Pompeii, Herculaneum and the Museo Archeologico in Naples. Deeply enamoured by what they saw in these historical places, and in the gardens of wealthy European homes, the British upper classes wished to have something similar in their own estates back home. Antique statues, urns and columns all found their way back to the British Isles.

The supply of original Roman and Greek antiquities was limited, however, so even as early as the 1700s the Italians were enjoying a flourishing trade in reproductions. Skilled Italian artisans copied the original designs in beautiful stone and marble, and many of these high quality reproductions ended up in grand British gardens. Local British stonemasons would, in turn, then carefully copy these ornaments.

Given huge budgets with which to work, eighteenth-century designers filled the gardens of the rich with ornaments from classical antiquity. The grounds of stately homes were stuffed with finials and obelisks, statues and urns, temples and giant fountains – all designed to carry the grandeur of the house into the garden. Such impressive classical items served as reminders of past civilizations and the perceived romance of times gone past. Architectural follies made a statement about the landowner: that he not only owned the land, but also dominated the landscape and had the power to shape nature at will.

By the time of the Industrial Revolution, grand ornaments and sculptures were well within the reach of the newly wealthy middle class. Keen to emulate the grand gardens of the English aristocracy, the Victorians bought reproduction marble, stone, lead, composition stone, terracotta and bronze ornaments from all over the world, including France, Italy and Japan. Imports from Asia led to a fascination with all things oriental, and it wasn't uncommon to see a Victorian English garden festooned with oriental trellis, *kasuge* (large stone lanterns) and wooden bridges.

At the same time, the technology became available to mass produce cast-iron garden furniture. Foundries such as Coalbrookdale, Carron and Falkirk all flourished supplying the huge demand for fancy iron seats, chairs and tables. The ease with which cast iron could be made into a variety of shapes and designs also allowed the Victorians to indulge their wildest excesses and create extravagant cast-iron fountains, gazebos, urns, birdbaths and many other garden ornaments.

Glasshouses became popular during the Victorian age, inspired, in part, by the greatest glasshouse of all at the Great Exhibition of 1851 – the Crystal Palace. Used to grow unusual, exotic and often delicate plants, the glasshouse became a common feature in wealthier homes, as did charming wooden summerhouses. Few of these original structures have managed to survive, and a restored or intact example can fetch a high price at auction.

Nowadays, alongside the thriving market in antique garden components – statues, urns, birdbaths and so on – you find an equally lively trade in other, more unusual types of garden ornament. Garden designers are creating exciting, original schemes using a wide variety of architectural elements: old agricultural machinery, columns, iron beds, stone troughs, staddle stones (stone supports for wooden barns that look like giant mushrooms), garden tools, rainwater hoppers, lead pumps and sinks. The most sought-after items are always the ones with the best patina or quirkiest touch, something that adds a sense of history, narrative or drama to the garden.

SOURCES

Practically every salvage yard will stock some piece of garden salvage, but there are fantastic specialist dealers who trade exclusively in this type of architectural antique. There you'll find all sorts of classical garden goodies – distressed statuary and mythical beasts – alongside other types of outdoor ornamentation including mossy benches, gazebos, gates, fountains, sundials, weather vanes, chimney pots, urns, lampposts and butler's sinks.

Eighteenth- or nineteenth-century statuary is expensive – often ending up in London showrooms and top auction houses to be snapped up by the American market. These pieces go for thousands of pounds but, like so many high quality antiques, represent an investment for the future. Good examples can fetch well over £10,000. Believe it or not, you can actually still buy genuine ancient artefacts – whole Roman or Greek statues are rare and very expensive, but you can pick up a classical marble head for just a few thousand pounds. You probably won't find these in salvage yards; you'll have to go to a specialist antiquities dealer. Just make sure they are a Member of the Antiquities Dealers Association and abide by the standards of its code of conduct.

You can also expect to pay high prices for early cast-iron benches, such as those made by the English Coalbrookdale foundry or the American producer J. L. Mott & Co. They are highly collectable and a good example can be worth thousands of pounds. Unfortunately, the demand has led to a lot of cheap aluminium fakes on the market. It can be difficult to tell, but a genuine cast-iron bench should be very heavy and take at least two people to lift. You can also take a magnet with you to tell whether a bench is aluminium or cast iron (a magnet will stick to cast iron but not to aluminium). Above all, the price should be an indication of what you're getting.

Bronze garden ornaments often have a high value, while Far Eastern modern copies will be worth much less. It can be difficult for a novice to tell the difference, but the detailing should be crisp and good quality. Resin bronze replicas are also common. To tell the difference, give the object a quick tap: bronze will resonate like a bell, while resin will make a dull knock.

Lead ornaments are very collectable, even though the majority of lead garden objects available today only date to the last century. As with all garden ornaments, the quality of the decoration and casting make or break the value. Leadwork produced by the Bromsgrove Guild is always high quality, and therefore highly sought after, but you can pick up less well-made late twentieth-century pieces for just a few hundred pounds.

If you don't want to spend a fortune, there are also real bargains to be had in a garden salvage yard. Cast- and wrought-iron gates often represent good value and can be cheaper than modern reproductions, even after restoration. Wooden gates can also be picked up fairly cheaply. Flaking paint is nothing to worry about (in fact, it can add to a gate's value) but keep an eye out for any areas of rot. Gates are relatively inexpensive because they're readily available but not always easy to fit into a new space. It's easier to plan an opening around an existing pair of gates than try to adapt them to fit. Side gates, however, are pretty much a standard size and shouldn't require alteration.

Twentieth-century composition stone urns are much more affordable than their carved stone counterparts and will soon weather to fit in with your garden. They won't increase in value, however, at the same rate as a solid stone urn. Saying that, some of the early composition stone examples (late nineteenth-century and earlier) do fetch surprisingly high amounts at auction.

Delicate wirework benches and gazebos are now cheaper than they were at the beginning of the 1990s, due to an influx of modern wirework flooding the market. Late nineteenth- and twentieth-century pub tables – small round tables with a cast-iron base and a wooden, marble or cast-iron top – are also inexpensive but perfect as a sturdy table for two in the garden. You'll generally pay more for a table with a cast-iron top than a wooden or marble top.

Antique lead hoppers (water funnels usually found at the top of drainpipes) are also good value, easy to find and surprisingly decorative. Many are stamped with dates or fancy motifs and make lovely planters when fixed to the wall. People also convert them to water features (although not to drink out of, lead being toxic). Victorian chimney pots are a real bargain and easy to find – use them as pedestals, planters or forcing pots. Get four the same size and you've got sturdy legs for an outdoor table.

Gardening tools and implements – forks, watering cans, trowels, rakes – are also cheap to buy and make fantastic 'folk art' decorations, even if they're too fragile to use. Hang them on the wall like pieces of art. Antique terracotta plant pots look much more attractive than their modern counterparts and can be picked up in batches for next to nothing. Look for ones with unusual detailing or a nice patina.

If you're not too bothered by a bit of damage, very weathered, cracked or chipped stone ornaments can be surprisingly cheap. They're not great investments but they can really add character to a garden and you don't have to worry too much about keeping them pristine. Mixing and matching is also a good way to get a bargain; think about four different garden chairs with a table instead of trying to find the perfect set.

If you can't find or can't afford the antique you want, consider a well-made reproduction. Garden centres have given repro classical statues a bad name, selling ghastly copies made from concrete, but you can get high quality reconstituted marble and stone statuary if you know where to look. Firms such as Cranbourne Stone, Haddonstone and Chilstone, for example, reproduce historic garden statuary found in the grounds of some of the country's most famous stately homes. Coalbrookdale produce good quality reproductions of their famous cast-iron benches, while the Bulbeck Foundry uses traditional techniques to produce copies of antique leadwork from the seventeenth, eighteenth and nineteenth centuries.

TRANSPORTATION

Before you spend any money on a piece of garden salvage, you need to think carefully about how you'll get it home. Most stonework is enormously heavy and easily damaged in transport. Ideally, you should arrange for the vendor to deliver your new purchase with a guarantee that it will arrive in the same condition it left the yard or auction house. Lead troughs, for example, require careful handling as the metal is soft and easily damaged. You'll also need to make a quick decision about where it's going to go in the garden – once it's delivered, a huge piece of sculpture or stonework will be difficult to move again.

If you want to put your garden ornament on a plinth or wooden structure such as decking, make sure it can support its weight. Equally, if you place your statue or urn on the ground, ensure that little hands won't be able to push it over.

RESTORATION AND CLEANING

To get the best from garden antiques it's important to know how to care for them:

CLEANING

The worst thing you can do to a statue or similar item is over-clean it. Pressure-washing a garden antique to remove layers of moss and lichen is almost certainly going to knock value off it and may even damage it. If a limestone or sandstone piece has a powdery or flaking surface, definitely don't clean it yourself – take advice from a conservator.

Dolf Sweerts de Landas suggests that 'masterly inactivity is the best thing' when it comes to antique statuary, but if you insist on cleaning a piece, only use warm soapy water and a soft brush. Start from the bottom and work your way up, don't let water run down the surface and hold a cloth to catch the drips. Rinse with clean water. Remember never to wash stone in the winter, when there is a strong risk of frost.

If you just need to remove bits of loose dirt, simply brush gently with a natural bristle brush, taking care with any delicate areas. Never use bleach to whiten stained marble as it can corrode the surface.

AGEING

If, on the other hand, you want to encourage lichen and moss to grow on your garden antiques, don't make the common mistake of using yoghurt. Plant food, bird droppings or manure are better choices for ageing statuary. These protein-rich sources feed the lichen and mosses and encourage them to grow, while yoghurt will only attract mould.

FROST

A bad spell of frost or ice can wreak havoc on stone, terracotta and marble. The National Trust suggests that, if you can't move these objects indoors for the winter, cover them with canvas or wooden covers to keep the cold at bay. Do this

before winter is in full swing and when the object is still dry. You should also empty any stone urns and troughs to reduce the chance of them being split by extreme cold weather.

Fountains, in particular, are vulnerable to the extremes of northern climates. When water freezes in a fountain it can burst the metal pipes and crack the stone bowl. If possible, it's best to leave the fountain running throughout the year. If a really cold snap is on its way, it may be prudent to drain the fountain or install some kind of heating device to prevent the running water from reaching freezing point.

DAMP

When you position your stone garden antique, take care to protect it from sources of damp and mould. Low-hanging tree branches may cause water to drip on to your stone sculpture and stain it, while overhanging ivy can encourage algae. If your statue is placed on the ground, prevent damp rising through the stone by placing a damp-proof membrane under the plinth or base.

REPAIRS

If you want to achieve the timeless, well-worn look that is so characteristic of salvage, the garden is one of the best places to indulge yourself. Damaged and broken antiques can be used just as they are and add a wonderful sense of history to your garden, as if a piece has been there since the garden was first created. A rusty garden gate, a statue missing an arm, or a cast-iron bench with peeling paint – these are typical motifs of an eccentric and interesting English garden, and can be used to great effect.

What you mustn't do is artificially distress a statue with paint and a sponge – it will never look like weathered stone and waterproof paints may even trap in moisture, damaging your ornament. You'll also knock significant value off the piece.

If you do want to make repairs, however, it's worth consulting an expert, especially for the more valuable pieces. Where repairs are required to a stone statue, for example, a master stonemason will be needed if you want the job done properly.

USING GARDEN SALVAGE

When it comes to design and decoration, your garden deserves just as much thought as the inside of your home. Depending on the look you want to achieve, a few well-chosen pieces of garden salvage can make a real impact, injecting a sense of playful drama. You could stick to classical themes, choosing just a few Romanesque ornaments and the odd draped muse hidden among the bushes to create the feel of faded gentrification. Or what about a Gothic garden, with a half-built folly, pointed arches and a grotesque gargoyle? A romantic garden could contain stone cherubs, a birdbath and a pretty summerhouse. Go for a French chateau theme – a simple fountain, a beautiful cast-iron railing and an exquisite wirework gazebo. Or how about a rustic cottage garden: stone troughs filled with flowers, a hand-pump water feature and stacks of terracotta long-toms?

Whatever you choose, here are a few pointers to help you get the best from your garden salvage:

- Large pieces of salvage or sculpture look great placed against a simple backdrop. Think about what could go behind your ornament: a dark dense hedge of yew or box, or a wall covered in ivy, for example. Frame it like a picture. See how it looks close up and at a distance.
- Most small pieces of sculpture work best on a plinth – not only does this prevent damp rising up through the statue or urn, but it also adds an extra sense of importance and dignity to the piece.
- If you've bought an expensive piece, don't clutter it with other smaller pieces. If, on the other hand, you want an eclectic, busy garden, choose smaller, inexpensive pieces such as terracotta plant pots and wooden planters, and pile them high.
- Don't be afraid to buy a really large piece of garden salvage. Items such as a wooden hand cart or a huge urn will create a fantastic talking point and a real focus for the garden.
- Don't stick slavishly to a certain era. A modern sculpture can look breathtaking in a traditional formal garden, while nothing is more romantic than a classical sculpture peeking out of a hectic, blowzy rose bush.
- Use religious statues and imagery with some degree of sensitivity: angels and gargoyles can be fun and frivolous, but some neighbours might

struggle to see the funny side of the Virgin Mary being used as a model for twinkly fairy lights.

- Sit your statues by the edge of a pool. The water will act as a mirror and the reflections can be captivating.
- If you're drawing your garden design from scratch, consider a 'splurge' piece as your starting point and build around it. A beautiful piece of garden salvage can be as captivating and eye-catching as hundreds of pounds' worth of planting.
- Think about using things that weren't originally intended for the garden. Have fun. A rowing boat sawn in half makes an unusual garden seat, butler's sinks are often used as planters and a nineteenth-century pew will create a comfy garden bench. A cast-iron bed will survive perfectly well outside and, covered with cushions and throws in the summer, will be a fantastic garden couch.
- In the winter, bring the outside in. Drag your urns into the living room and fill them with collectables or Christmas decorations. Bring the statue into the hall and let her greet your guests on their arrival. Chimney pots make fantastic umbrella or walking stick holders. Terracotta pots look gorgeous in the kitchen stuffed with cutlery or in the bathroom filled with soaps or toothbrushes. Metal café chairs can become impromptu seating and watering cans are the ideal containers for cut flowers.

SECURITY

One of the only downsides to garden salvage is that it's more easily stolen than salvage kept indoors. The value of garden antiques has rocketed in the last few decades and this fact hasn't gone unnoticed by the criminal fraternity. The scale of the thefts is growing and ranges from statues worth tens of thousands to stone troughs worth just a few hundred. Some of it gets sold off in this country, while much ends up abroad.

What kind of security measures you take depends on where you live and what you have in your garden; there's obviously a big difference between keeping a £20,000 statue and a £500 stone trough. If you love what you have on your lawn, however, chances are someone else will. So, even if the object is worth just a few hundred pounds it's probably a good idea to use some of the security measures described below. You may also find that insurance companies won't insure garden antiques unless they're satisfied you've taken adequate precautions.

TAKE PHOTOGRAPHS

Photographs of a stolen object greatly improve the chances of it being recovered by the police, should the worst happen. Use whatever photographic format you feel comfortable with – digital, 35mm, instant – this doesn't have to be an expensive exercise. Take a variety of shots of each object. Vary the angle and remember to focus on what makes the object unique and identifiable.

Indicate the object's size and dimensions by placing a metre rule next to it and show any distinguishing marks, repairs or maker's names. Take photographs from both back and front. Remember, a valuable object doesn't have to be one of financial worth – record any items that have sentimental value too.

WRITE A SHORT DESCRIPTION

This will be much easier to do now, rather than trying to remember the details of an object if it's been stolen. Make yourself an object description form and staple any relevant photographs to it. Keep these documents in a safe place. On the form try to include:

- What type of object it is: urn, statue, bench, sundial etc.
- What it's made from: stone, lead, marble, bronze
- Exact measurements in metric
- Any inscriptions and markings
- Distinguishing features: damage, repairs, manufacturing defects, unusual detailing
- Title: name of the urn or sculpture if it has one
- Maker: Coalbrookdale, Austin & Seeley, LEFCO, for example
- Date: even if it's just an approximation or century

At the same time, be aware of the limits of your house and contents insurance. You'll need to make sure you have adequate cover for any significant objects. List any items worth over £500 and tell your insurance company their exact value. Some brokers have special garden cover – this is probably better than your average high-street policy. Ensure you'll receive a cash settlement should anything be stolen – that way you'll have the choice about what you replace and where you buy it from.

SECURITY DEVICES

A number of companies specialize in garden ornament security. They may offer one or more of the following anti-theft and identification measures:

- **Screech alarm.** Also known as a 'lock alarm', this will emit a high-pitched noise and flashing lights if tampered with.
- **Garden sitter.** More sophisticated than a screech alarm, a 'garden sitter' alarm uses sensors to detect movement around an object. If movement is detected, a signal is sent to a receiver in your house alerting the owner that the object may be in the process of being stolen.
- **Anchors.** Different ways effectively to tie your garden antique to the ground – from heavy-duty stainless-steel rods and chains to thinner, more discreet cables.
- **Tags.** Microdots, ID tags and other tiny devices can be attached or inserted into your garden antique. Just like microchipping your pet, if your salvage goes walkabout and turns up elsewhere, an electronic tag reader will instantly identify the owner of an item.
- **Replicas.** Some people have replicas made to put outside the house and store the valuable antiques inside. This is probably only necessary for the larger estates and collections, where it's difficult to keep an eye on everything at once.

MANAGE THE RISK

There are also security measures you can take for yourself:

- Make it as difficult as possible for a thief to take something away. Use high walls, spiky hedges, fences, ditches and sturdy gates to make physical barriers.
- Don't put expensive garden salvage in the front garden – if possible, keep it to the back garden where it won't attract too much attention.
- Don't advertise your expensive salvage. Owners of impressive houses sometimes like to appear in interior design or lifestyle magazines and gangs of thieves are thought to study such magazines, looking for items to steal to order.
- A guard dog is often the best deterrent.

IF THE WORST SHOULD HAPPEN …

. . . Don't panic. Whatever you do, don't simply put it down to bad luck and not tell anyone. The authorities will want to know if something's been stolen and there might be a chance you'll get it returned if you act quickly.

- Contact your local police station and report the item as stolen.
- Check the Salvo website – it has been publishing theft alerts since 1990.
- Contact the Art Loss Register (www.artloss.com) who, for a small admin charge, will take a note of your stolen property and search auction houses and online sales to see if it reappears.
- Look in *Trace* magazine or www.trace.co.uk, a catalogue of stolen art, antiques and salvage with the corresponding police officer's details.

EXPERT TOP TIPS

Steve Tomlin of Minchinhampton Architectural Salvage: Garden salvage

- Always check the origins, history and background of an item.
- Check the general condition of the piece and look for bad repair jobs.
- Look for makers' trade marks and names.
- Establish whether the piece is made from natural or composition stone.
- When buying statues, look for fine detailing of hands, feet and facial expressions – all give clues to high quality.

IRONWORK

Before we take a brief glance at the history of ironwork, it's worth distinguishing between the two types dealt with here: *wrought iron* is worked into shape by hammering it on an anvil, while *cast iron* is hot liquid iron poured into moulds and left to set. Each type of iron has its own history and different uses.

Wrought iron is the oldest type of iron and has been used since, perhaps not surprisingly, the Iron Age. The blacksmith would hold an important role in the village or town, producing many of life's necessities – from everyday agricultural tools, horseshoes and ploughs to high prestige jewellery and armour.

Wrought iron was also important in the construction of early buildings – nails, spikes, door hinges, locks, handles, railings and brackets all had to be made by hand. One of the earliest surviving examples of early wrought ironwork in Britain comes from a door at St Helen's Church in Stillingfleet, East Yorkshire. Dating from the twelfth century, the door has two large iron hinge straps and beautifully simple wrought-iron representations of a ship, a man and a woman, a tree and an interlocked cross, possibly depicting the legend of the Holy Rood Tree from which the crucifixion cross was made.

The late seventeenth century saw a real revolution in techniques and designs in wrought-iron production. At this time, Britain was experiencing an influx of skilled craftsmen who were escaping religious persecution on the Continent. Among them was a Frenchman, Jean Tijou, who brought with him wonderful new styles of wrought ironwork, which included the lavish use of leaves, flowers, rosettes and figures. His beautiful wrought-iron gates and railings still adorn the grounds of Hampton Court Palace, which William and Mary commissioned soon after their accession to the throne in 1689.

Unlike wrought iron, cast iron was a rarer material throughout most of British history. Recent archaeological evidence suggests that people in England have been attempting to make cast iron since the Saxon times, but furnaces struggled to reach the temperature needed to turn iron molten. The introduction of the blast furnace in the fifteenth century helped cast iron to be produced reliably but the huge amounts of charcoal needed to fuel the furnaces made the process expensive. Only in 1709, when English iron maker Abraham Darby substituted coke for charcoal, was the production of abundant and cheap cast iron made possible.

At first, cast iron was used mainly for ornamental purposes, but in 1779 the first major structural use of cast iron came in the form of the Iron Bridge in Shropshire. The Victorians took to the new metal immediately. It was cheaper to produce than wrought iron and the casting process allowed hundreds of identical components to be made in one go – perfect for large-scale building projects and industrial machinery, as well as domestic items including gates, railings, stoves, rainwater goods, fireplaces, baths and garden furniture. But cast iron was also easy to fashion into wonderfully complicated designs and motifs; buildings such as Paxton's Crystal Palace and the Eiffel Tower in Paris proved that cast iron was not only robust as a structural material, but it could also be used with great sensitivity, delicacy and imagination.

SOURCES

The majority of ironwork that finds its way into salvage yards is Victorian and Edwardian cast iron. These pieces come from all kinds of buildings and structures: factories, churches, bridges, railway stations, streets and houses. You'll also find pieces of modern reproduction cast iron for sale as well as pieces of antique wrought iron – especially smaller items such as door furniture, gates and railings.

Look out for the work of famous foundries such as Carron. Founded in 1759, it was one of the greatest ironworks in Europe, famed for its cannons, post boxes and red telephone boxes, among other things. Some more popular names to look out for include Coalbrookdale (benches, seats and stoves), Gaskell & Chambers (tables), Kenrick (domestic items and door furniture), James Yates, Falkirk, and Handyside.

When you are looking at a piece of ironwork, condition is key. Check the joints and corners – these are especially vulnerable to rust and decay. Cast iron is not always easy to restore, so make sure any pieces are complete before you buy. This applies particularly to fireplaces. Thick layers of paint may conceal hidden damage or rust; it's always a risk buying a piece unrestored. If you're spending a lot of money on a large piece of cast iron – say a spiral staircase – make sure restoration costs are part of the deal. Otherwise, you could find yourself stripping back decades of paint to reveal a severely corroded and potentially useless architectural feature. On the whole, however, pieces of reclaimed ironwork will be in good condition and just need a little cleaning and the odd spot of rust removal. Antique ironwork was built to last and most pieces still look as impressive as the day they were made.

N.B. One important point to remember if you're planning to scrabble around a salvage yard in search of old metal – tetanus. You will have probably had a vaccination in your childhood, but you need to keep up with your ten-year booster vaccinations, just to be on the safe side.

RESTORATION AND CLEANING

CLEANING

Water is the number-one enemy of iron, causing it to rust and decay, so the top priority for any piece of antique ironwork is to keep it clean, sealed and dry. If your piece of salvaged ironwork just needs a good clean, remove any dirt with methylated spirit on a soft cloth. Don't use water as it can encourage rust to form, damaging the surface.

You can combat the early stages of rust – surface rust – with a commercial rust remover and a pad of super-fine steel wool. Once you've removed the surface rust you'll need to think about a sealant, to protect from any further moisture penetration. What finish you use will depend on where you plan to keep your ironwork.

If you want painted exterior ironwork, apply a rust protector, a primer and then paint with a suitable exterior finish: apply two or three thin coats of paint rather than one thick coat. If you don't want a painted finish you can also finish exterior ironwork with boiled linseed oil. Apply with a clean rag, working the oil into the metal. Leave for ten minutes to let the linseed oil soak in and then remove the excess.

For interior ironwork such as cast-iron fireplaces, you can use an appropriate polish like Liberon Iron Paste or Zebo polish and buff to a sheen with a soft cloth.

1857
1857

For interior wrought ironwork, *The English Heritage Technical Handbook* suggests a number of different oils and waxes, including a combination of beeswax and boiled linseed oil, as well as more old-fashioned treatments such as lamp black and goose grease.

If you need to remove any layers of old paint, use an appropriate paint stripper. You can also get ironwork shot-blasted; this is fine for cast iron and modern steel, but don't use it on delicate wrought iron. Wrought iron has a thin outer surface called 'mill scale', which provides an important natural layer of protection against decay. Grit-blasting can damage this layer of mill scale and also remove any unique characteristics, such as original tool marks or very fine detailing, from the surface. If you must strip old paint from real wrought iron, use a chemical stripper.

REPAIRS

If rust has had a chance to take hold, you may find that more significant action needs to be taken. Cast iron can be difficult to repair and may need specialist skills such as gas fusion welding, metal stitching and electric arc welding. There are specialist cast-iron welders in the UK, such as www.cast-iron-repair.com, but your first port of call should be the salvage yard. Ask if they have an in-house repair service or know someone who could help. Repairs won't be cheap, however, so factor this in to your budget from the start.

Wrought iron is easier to restore than cast iron, but the work still needs to be carried out by a traditional blacksmith or wrought-iron restorer. Some minor repairs can be performed *in situ*, but most work will need to be done in a workshop, where special tools and materials are available on hand. Repairs to wrought iron are best done with the same material and techniques used originally – in other words, repair wrought iron with wrought iron. Steel can be welded to wrought iron but it isn't really suitable for restoration work because the properties of the two materials are so different. Steel will rust more quickly than wrought iron and any repairs made in steel will lack the character of early, hand-forged iron. It's true that steel will be cheaper to use than wrought iron, but either way the largest part of your restoration costs will come from labour, not materials. Get a few estimates before you commit.

USING IRONWORK SALVAGE

Iron gates, balcony rails and railings are some of the most common items in the salvage yard. Old gates can be restored or altered, but it can be cheaper in the long run just to move the gateposts. It's worth knowing that side gates haven't changed in size since the Victorian era, so they're easy to accommodate into a new space. Very ornate gates and railings are pieces of art in themselves and can look fantastic as interior features even in the most contemporary of settings. Broken gates and sets of railings don't have to be consigned to the scrap heap: use them as trellis for climbing plants or as garden edging.

Iron heating vents have been around since Victorian times, when the first forced air heating systems were introduced. Common in America, these large metal grates provide perfect table tops, paving, and ornate doormats. You can also use cast-iron ventilation grilles to conceal and enhance a modern radiator.

Old cast-iron or wrought-iron door and window furniture can be surprisingly ornate underneath all the layers of paint that have accumulated over the years. They are often better quality than modern hardware and easily incorporated into new buildings. And don't forget iron brackets – these can easily be used to hold up a wooden or glass shelf.

You can also find iron banisters, guttering, radiators, stoves, fireplaces, lampposts, cooking ranges, benches, seats, firebaskets and fire dogs, statues, baths, window frames, floor grilles, urns, cinema seats, posts, columns, shop signs, tables, toilet cisterns, beds, brackets, telephone boxes and police boxes.

EXPERT TOP TIPS

Andy Hayward of Original Architectural Antiques: Ironwork

- Check the rust has not gone too far, especially in the corners.
- Don't worry about build-up of paint on cast iron; shot-blasting can bring back detail.
- Use a good red oxide primer to prevent rust coming through.
- Old wrought and cast iron will outlive any modern-day ironwork.
- Remember that cast iron is very brittle.
- Make sure pieces are complete as you can't bend cast iron and therefore it is difficult to repair.

KITCHENS AND KITCHENALIA

Before the eighteenth century, cooking and kitchen appliances had remained virtually unchanged for a thousand years. Little separated the technology available to rich or poor; both had to rely on open fires for cooking, no running water and no electricity.

The only difference lay in the scale of operations. A nobleman's house would have a huge kitchen, bursting with kitchen staff and servants. Centred around a large preparation table, you would find wooden dressers, cupboards and other free-standing storage areas stuffed to the brim with pots, pans and other cooking and eating utensils. This room was purely utilitarian in function – large, almost industrial in scale and purpose, and a million miles away from the calm serenity of the master's dining room.

In a peasant's cottage the situation was very different. It wasn't unusual for an entire family to live in just one or two rooms, so the notion of a separate kitchen simply didn't exist. Cooking would have been performed in a large pot over the same fire that warmed the house, lit the room and provided all the hot water for the family. This living/kitchen room may have contained a table and a small dresser, but by and large there was no special furniture designed for food preparation. Meals were eaten sitting on a bench or stool close to the fire for warmth.

The first significant change was the introduction of the cooking stove. Before the 1700s all meals were cooked on open fires. In wealthier homes, servants used hooks and pulleys to raise or lower cauldrons over the hearth, controlling the amount of heat supplied. Clay, brick and stove ovens were generally restricted to bakers and the very rich.

Between 1770 and 1820 home owners began to incorporate cast-iron stoves into their kitchens. The early examples were difficult to control and lacked thermostats – cooks had to use a lot of guesswork when it came to judging temperature and cooking times. Iron hotplates were also added to stoves, which meant kettles and pots could be safely heated on a flat surface.

The Victorians went even further with kitchen technology. Keen diners and lovers of extravagant food, the Victorians channelled their inventiveness into the kitchen. The middle of the nineteenth century saw the introduction of oil- and gas-fired stoves (even the most humble of terraced houses would be built with a cast-iron range incorporating an oven, hotplate and register-grate open fire). By the 1890s the Victorians had experimented with the first electrical stoves.

The Victorians also loved their gadgets and it was at this time that many of the utensils we take for granted appeared for the first time, including egg whisks, cherry stoners and apple peelers. In fact, the Victorian kitchen would have been stuffed with labour-saving devices, all designed to help with the mass of work that had to be done to cater for a large middle-class family. Typical appliances from that time also include knife-sharpening machines, lemon squeezers, parsley choppers, sugar snips, nutmeg graters and early hand-turned food processors.

Even the dishwasher was an invention of the Victorian age. The American Joel Houghton invented the first prototype in 1850, made from wood and operated by hand. The first automatic dishwasher followed in 1886, from an enterprising rich socialite called Josephine Cochrane, who had had enough of her servants breaking the crockery. Her idea was a smash hit and soon she was taking orders from the restaurant industry and hotels. Her famous company KitchenAid is still going strong today.

Throughout the Victorian era upper- and middle-class families would still enjoy the privileges of domestic help. Kitchens would be placed in the basements of town houses or in the rear of the house, distinctly separate from the rest of the dwelling. Largely free from ornamentation, these kitchens were dowdy and plain in comparison to the over-decorated excesses of other rooms.

By the end of the First World War, however, the supply of domestic staff started to dry up. The changing social and economic climate meant that many people didn't want, or need, to take employment as house servants. Women were encouraged into professions previously dominated by men, such as office and factory work, and could often enjoy better wages and working conditions. The super-rich would always have their share of hired help, but the middle classes would have to swallow their pride and start cooking for themselves. The age of the kitchen as an informal cooking and eating area had begun.

Another change came with the introduction of the fridge. Before the 1830s, perishable food was kept cool with an icebox – an insulated container that held large chunks of ice. Demand for ice was so high that in the 1860s a huge well was built underneath central London to contain 700 tonnes of ice. Despite the fact that this volume could stay frozen for months, individual iceboxes were messy and inconvenient, often needing refilling on a weekly basis. The invention of ice-making technology was a relief. In 1867 a refrigerated rail car was introduced and by the end of the century the first commercial refrigerators became available.

The Americans were thrilled. After the First World War, many US households replaced their antiquated iceboxes with shiny new fridges, some of which even had a small freezer compartment. The British, however, took a bit more persuading. Despite the fact that early prototypes had been around since the 1890s and the first British Frigidaire had been sold in 1924, even in the mid 1950s still only eight per cent of English homes contained a fridge – compared to a whopping eighty per cent of American households.

The 1920s saw the introduction of the first Agas. Invented by Swedish engineer Nils Gustav Dalen, the Amalgamated Gas Accumulator Company patented his design and put the first Aga stoves into production in 1922. These popular cooking stoves – which run on oil, gas, solid fuel and now electricity – produce a radiant heat that keeps the kitchen warm, as well as providing heating and hot water for the rest of the house, should you wish.

After the Second World War, Britain emerged from a time of austerity into the optimistic and colourful 1950s. Modern designs for the home were inspired by the advent of space exploration and looked to the future with their sleek appearance, modern stylish curves and bright colours. Two of the most famous kitchen companies to come out of this era were English Rose and Boulton Paul. During the war aluminium had been stockpiled for the aeroplane industry. When the need for war planes diminished with the end of the conflict, both factories turned their hands to fitted kitchens using the latest technology and skills from aircraft manufacture.

Both English Rose and Boulton Paul kitchens were originally painted (often cream) with brightly coloured handles and worktops. Their sweeping designs and graceful curves are an instant way to add retro chic to your house without compromising on quality or durability. Some salvage hunters prefer to strip the paint away, leaving gleaming polished aluminium surfaces that don't look out of place in an ultra-contemporary kitchen. Either way, it's pretty exciting to think that the chaps who made your Boulton Paul kitchen also built the Defiant fighter planes that took part in the Battle of Britain.

SOURCES

Free-standing kitchen units are now in great demand. Butcher's blocks, bread cupboards, large wooden tables, glazed cupboards, shelves, plate racks, dressers, corner cupboards – these are the must-have constituents of any modern country kitchen and good value at most salvage yards. Butcher's blocks are particularly lovely pieces of furniture. Usually hefty chucks of wood, cut across the grain, these have distinctive and hugely appealing patterns of wear and tear and cut marks. They also transfer readily into any modern kitchen and will happily sit alongside stainless-steel cookers and high-tech appliances. Don't be tempted to sand down the surface of your butcher's block. The cuts and knocks are part of its charm and certainly add to its value.

Sinks are also commonplace in salvage yards. Depending on their age and provenance, they can be made from stone, lead-lined timber, tin, wood, earth-enware or ceramic. They may come with cast-iron legs or be set on top of a timber worktop. Porcelain-glazed sinks such as the chunky white Belfast sinks and more shallow gamekeeper's sinks are still reasonably priced, but watch out for chips and cracks, as they can be prohibitively expensive to repair. Old laboratories and hospitals can yield fantastic porcelain sinks, often double or triple your normal sink size. Look out for Edinburgh sinks, which have an attractive bowed front. Some old sinks don't have overflows, so check that they are fit for their intended purpose before you buy. Look out for vintage draining boards and taps to complement your sink.

Second-hand stoves also appear from time to time. Agas, Rayburns and other vintage stoves are synonymous with the country kitchen, but make sure you have enough space to accommodate one before you buy. They are also extremely heavy, so you'll probably need to pay to get it from the yard to your home. If the stove needs installing by a professional or totally restoring, factor that into the total cost as well.

Look out for vintage kitchen accessories – rolling pins, drying racks, whisks, display boards, breadbins, tins, meat and pie safes, egg baskets, spice boxes, meat grinders and colanders. Early and mid twentieth-century weighing scales can be picked up for next to nothing – just make sure the mechanism still works as finding replacement parts isn't always easy. (If you have an old set of Avery scales, the Avery Historical Museum in Smethwick, West Midlands is a really useful source of information and advice.)

OXO
Bread

RESTORATION AND CLEANING

RE-ENAMELLING BELFAST SINKS

Old chipped Belfast sinks can be re-enamelled by a bath restorer, but this is often as expensive as buying a new one. Weigh up the costs before you commit. It might be cheaper to get a few sinks re-enamelled in one go. If you can't afford a full restoration, an old kitchen sink is still jolly useful out in the garden, either as a utility sink or as a planter. Antique brass taps are still good value; if you can't find a set that you like, ask the salvage yard if they can strip the chromium plate from a set of chrome taps instead. Don't be tempted to varnish your brass taps – just give them a polish now and again.

RESTORING RETRO FRIDGES

Many salvage customers are willing to forgo all the convenience of modern fridges for the soft, rounded look of genuine retro appliances. Early to mid twentieth-century fridges are a labour of love to restore, so you need to know what you're getting yourself into. The cost of restored appliances can often exceed the price of a new model, and there's a good reason for this. To restore a vintage fridge, it needs to be taken apart completely and stripped of all the old insulation and wiring. Any internal repairs will need to be made and the outside will have to be primed and finished with enamel. The whole fridge then gets new insulation, new coolant, new wiring and an overhaul of the mechanical system. Not a quick job. Once they're restored, however, they're absolutely beautiful.

EARLY FITTED KITCHENS

Some salvage companies specialize in restoring vintage fitted kitchens. English Rose and Boulton Paul units are the most highly sought after (see page 145) and you can choose to have them powder-coat painted (which gives an ultra smooth finish) or get a specialist salvage dealer to strip them back to their gleaming aluminium core. The result, when polished, is absolutely gorgeous and totally contemporary. If you want the genuine retro look, team the painted units with that 1950s kitchen essential, the Formica worktop.

STOVES AND RANGES

There are specialist companies who offer reconditioned and second-hand Agas and Rayburns; these are usually cheaper than buying brand new and will add a little retrospective charm to your kitchen. If you've already bought an antique stove, it may need a total overhaul. Companies like Twyford Cookers offer a full Aga renovation service, including a fuel change (if you wanted to change from solid fuel to gas, for example), re-enamelling, shot-blasting of cast-iron internal components and replacement parts. For those people lucky enough to have a genuine Victorian cast-iron range, there are a few dedicated restorers dotted up and down the UK including the Yorkshire Range Company who will be able to get it back to full working glory. Bear in mind, however, that restoring a range will be expensive – be prepared to drive a hard bargain for an unrestored example or you'll end up spending a fortune overall. N.B. Any gas appliance will have to be fitted by a CORGI-registered tradesman – never fit a gas cooker yourself.

WORKTOPS

Salvage yards are often an inexpensive source of marble – you may be able to find pieces large enough to turn into a kitchen worktop. To tidy up a marble slab, use a pH neutral cleaner or, if the surface is very stained, try a poulticing paste such as Sepiolite to draw out the stain. Fill fine cracks with a suitable polyester, epoxy or cement-based filler coloured to match the stone.

You may also find a good selection of wood worktops. Mahogany, teak, maple and oak make wonderfully durable work surfaces, but the Victorians also used pine (which they regularly scrubbed, sanded and continually replaced). Some of these worktops may come from school labs, hospitals or morgues, so it pays not to be too squeamish about your salvage.

TILES

Ceramic tiles are one of the easiest salvage materials to use in your kitchen. Hardwearing, waterproof and easy to clean, they make the perfect material for kitchen splashbacks. The choice of colours and patterns is almost limitless: you can either go for a quirky mix-and-match effect or stick to the same design. One of the benefits of the eclectic approach is that if an antique tile gets damaged in the future, it might be impossible to find another to match it. To insure against this happening, don't use all your reclaimed antique tiles for your kitchen project – keep a few back in case of future breakages. If you don't mind broken tiles, however, there's an endless supply at salvage yards and they make the perfect, cheap mosaic material for resourceful renovators. If you've found a set of tiles that you really like, but they don't cover the area you need, consider incorporating them into a larger design.

USING KITCHEN SALVAGE

If you want to re-create an authentic Victorian kitchen, remember that fully fitted kitchens are a mid twentieth-century idea. Victorian kitchens would have used free-standing cupboards and tables, although built-in kitchen dressers were a common feature. Keep the space and furniture as sparse and functional as possible – a scrubbed wooden table, plain chairs, stone or tiled floors and a Belfast sink will look more authentic than a fully fitted hand-made kitchen.

The Victorian kitchen would also have been quite dull in terms of decoration – the cheerful fabrics and colourful tiles of 'antique-style' kitchens are a modern invention. But you don't have to be purist about things. Painting free-standing units can break up the monotony of too much bare wood and bring a bit of colour into the kitchen. If you find free-standing units that have genuine distressed paintwork, resist the urge to repaint, and use the aged, well-worn effect to its advantage. Distressed paint finishes are also user-friendly – you don't constantly have to worry about knocks and scrapes.

Mix it up. Antique chunky wood furniture and ceramic sinks will sit perfectly well with stainless-steel appliances and bright colours. Even just one piece of reclaimed salvage will soften up an otherwise brand-new cookie-cutter kitchen. Mix up the textures too: wood, glass, linoleum, stone, brick, metals. Have fun. Be inventive. Find outsized or undersized pieces for your kitchen.

If you want to play around with styles and eras, don't confine yourself to strictly domestic items. Industrial cookers can be picked up cheaply and fit in well with the current fashion for large, professional-size stoves. Granite work-tops will enhance most kitchens, as will hardwood counters from shops, schools, labs and factories. Canteen kitchen tables and factory assembly tables instantly solve the problem of large-scale entertaining, while church pews and chairs make fantastically chic budget seating. If you want to make a breakfast bar, old classroom stools are usually very cheap and can be repainted to suit your décor.

A selection of brightly coloured old Victorian tiles makes a cheerful splash-back. Laboratory and hospital sinks look very like Belfast sinks, but cost less and will often have interesting lever taps; just make sure they've been cleaned thoroughly before you install them – you don't want any nasty surprises!

Salvage from food and drinks companies can be used to great effect in the kitchen and include large copper vats, mixing bowls, soup pots, trays, trolleys, cake stands, racks, vending machines, rice steamers, butter churns, weights and measures, and sweet jars. Woven baskets from the woollen industry make perfect kitchen storage, while simple wooden church cupboards can easily pass for Shaker-style kitchen units. Museum display cases, metal hospital cupboards, mechanic's drawers and apothecary drawers also make fantastic additions to an eclectic kitchen. Use old window shutters or small glazed window frames as kitchen cupboard doors. If a salvaged cupboard door is missing its central panel, simply tack chicken wire on the back of the frame and call it a rustic 'food safe'.

To give your kitchen a 1950s retro feel, look for furniture and appliances with soft curves and pastel colours such as cream, mint, pink and light blue. Black and white chequerboard lino flooring will give your kitchen the American diner feel, as will polished chrome accessories. If you can find any salvage from a real 1950s diner, snap it up – it transfers really well into a modern kitchen and represents a smart investment for the future. Look for 1950s coffee counters, tiles, swivel chairs, juke boxes, clocks, radios, advertising signs, dining tables and chairs, and retro fridges. If you don't want to hunt around for retro appliances, modern up-to-the-minute chrome and stainless-steel fridges and cookers actually work perfectly well in a retro kitchen.

EXPERT TOP TIPS

Rod Donaldson of Source: Kitchen salvage

- Make sure you like what you're buying – you'll have to look at it every day.
- If the kitchen is not in perfect condition, make sure you're up to the job of restoring it.
- Make sure everything works or, if not, is replaceable; door handles and hinges are virtually impossible to find.
- Buy more than you need – you can always sell the excess afterwards.
- Be prepared to adjust rooms to fit pieces.

LIGHTS AND CHANDELIERS

Throughout most of history, people managed to light their houses without gas or electricity. Until the 1800s rich and poor families alike relied almost exclusively on candles for illumination: expensive, sweet-smelling beeswax for the wealthier households and malodorous tallow (animal fat) candles for the more modest homes. But candles were also expensive, so even in prosperous households they were rationed; a single candle would be carried to light the way from one room to another.

Even though the energy source was primitive, however, the designs for light fittings were surprisingly glamorous. Well-to-do houses might have a fine selection of ornate candleholders, wall sconces, candelabra and chandeliers made from brass, glass and silver. More modest homes would have had to make do with tin, wood and iron. The Georgians were especially fond of these types of decoration and the eighteenth century saw a proliferation of elegant candlesticks, lanterns, elaborate wall sconces and exquisite Adam crystal chandeliers. The latter were not only magnificent to look at, but the numerous glass droplets and beads served a useful function, reflecting and refracting light back into the room. Keeping the many candles on a chandelier alight, however, was an expensive and time-consuming task – one that would only be performed for special events such as weddings and very important guests.

Oil was also used as an early light source and people experimented with many different varieties including olive, fish, vegetable, nut and whale oil. A breakthrough came in 1784 when a rapeseed oil lamp (colza) was invented by Aimé Argand; his clever design produced a brighter light without any of the unpleasant smoke and strong odour associated with animal fat candles and other types of oil.

Unlike candles, which just need a candleholder, the colza lamps required a reservoir of fuel and a burner, as well as a glass shade to soften the harsh light. A variety of different designs soon emerged for the colza lamp: some of the smarter versions took their inspiration from classical urns, while others were turned into wall lights and reading lamps. The new oil lamps were cheap, easy to use and odourless, but despite all these benefits candles still remained the predominant method of lighting for both rich and poor homes until well into the nineteenth century.

The Victorians also had success with new lighting methods. Mineral oil lamps became available in the 1800s, including paraffin lamps, but it was the introduction of gas lights in the latter half of the century that altered how we think about lighting today.

Experiments with gas had been taking place as early as the 1680s, but it was in the mid 1850s that domestic piped gas became widely available in Britain. Unlike candles and oil lamps, gas light fittings needed to be attached to the mains gas supply. Light fittings became fixed permanently in place on walls and in the middle of ceilings, as they are today.

The early attempts at gas lighting were hampered by poor gas pressure and the corrosion of the metal burner, and it wasn't until the introduction of the gas mantle in 1886 that gas lighting really took off. (A mantle is the part of the lamp that creates light from heat, not from the flame, and allows a lamp to burn much more brightly.) People were also suspicious of gas to start with, and it wasn't until major institutions like the House of Commons incorporated new gas lighting that public perceptions began to change. Different types of gas light then illuminated the Victorian home, including the ingenious rise-and-fall gasolier – a light attached to the ceiling which could be raised and lowered as needed. The Victorians also experimented with stained-, etched- and leaded-glass gas lights, and you would often find a pair of gas lights mounted on the wall as part of the overmantel mirror.

Gas lighting was a great improvement on candles and oil lamps but it was the invention of electric lighting that truly revolutionized the home. The introduction of Thomas Edison's incandescent light bulb in 1879 promised a whole new world of illumination at the flick of a switch. Oil and gas lighting had cost about twice as much as candles, but electric lighting offered a practical and cheap alternative to all three. Electric lighting also had the benefit of being cleaner, easier and more convenient than any previous fuel source, and old gas lights could easily be converted to new electric supplies. The gas industry was mortified and retaliated with propaganda about the potential health hazards

of electricity. In desperation, it also brought out new designs for gas lamps, such as the inverted mantle.

Despite the life-changing possibilities of electric light, the earliest light bulbs produced a disappointing dim glow and were notoriously unreliable. Electricity was changing the face of cities at the turn of the century, but it still remained a minority energy source for most households. Only after the First World War did technical developments radically improve the quality and availability of domestic electric lighting, creating an opportunity for new lamp designs and light fittings.

Notable twentieth-century lighting designs include the organic shapes and sinuous flowing lines in Art Nouveau lamps by companies such as Tiffany and Liberty. Art Deco lighting is often characterized by figural lamps (female figures holding the ball of the lamp), 'flycatcher' ceiling shades and early chrome lights. You can also find exciting examples of retro lights from the 1950s, 1960s and 1970s with their bold colours and space-age designs. Look out especially for Danish and Scandinavian examples by companies like Nordisk Solar Compagni, whose lighting was designed by top architects of the day including Jørn Utzon, creator of the Sydney Opera House.

SOURCES

Most salvage yards will have a small selection of reclaimed lighting, but if you have a specific item in mind or want to see a larger selection, consider the benefits of the Internet. At sites such as Salvo and eBay you'll find a wealth of lights, lamps and chandeliers from all different periods and buildings. It's not all domestic lighting either; items such as Victorian street lamps, theatre lights, wartime searchlights, Tiffany Bell hotel chandeliers, 1950s tri-lights, railway lamps, hospital spotlights, Bakelite switches, 1960s brushed aluminium ceiling lights, restaurant fluorescent lighting, dentist's lights, neon motel signs, brass ship lamps, cinema wall sconces and Art Deco leaded glass ceiling lights are all available if you know where to look.

Dealers who specialize in antique lighting also exist and you'll find a bounty of reclaimed lights and lamps from the Victorian and Edwardian eras, including Arts and Crafts, Art Nouveau and Art Deco designs, as well as mid twentieth-century examples. Christopher Wray, for instance, started his huge lighting business by selling antique oil lamps. The King's Road branch in London has a section devoted to antique light fittings of all kinds, from large pendants and wall lights to oil lamps (antique oil lamps for sale must have a fuel label, e.g. paraffin, kerosene, petrol). LASSCO and Bygones in Canterbury also have a good selection.

Look out for table and ceiling lights, lanterns, sconces and standard lamps. Dig around for lights made with different materials: bronze, steel, aluminium, glass, wood, silver, plastic and enamel. If at all possible, buy originals – many reproductions simply don't have the same quality of manufacture and design.

RESTORATION AND CLEANING

Although the energy sources have changed over the centuries, designs of lights and lamps have stayed remarkably similar. This makes conversion from one type of energy source to another fairly straightforward, e.g. changing a chandelier from candle power to electricity.

Unless your salvage light fitting is very recent (1980s onwards) it will probably need adapting to modern standards; with lights this usually means replacing the bulb holder and flex and adding an earth. Existing old light switches will also need to be fitted with a new earth wire.

Old wiring is a common cause of electrocution and constitutes a serious fire hazard in the home. The rubber-coated electrical cables that were used in houses up to the 1950s, for example, are liable to perish, and introducing them to a new house puts you at risk from shocks and short circuits should the exposed wires make contact.

The salvage yard where you bought your light may offer to rewire it for you or know an electrician who could help. Alternatively, you can just contact a local lighting shop and ask if they have a rewiring service. Either way, unless you're a very competent DIYer, the rewiring of old electric lights is best left to qualified electricians who work to high safety standards. Remember, ideas of electrical safety were very different in Victorian times – you won't be able to use a wall-mounted light switch in your bathroom, for example, even if that was its original use.

Victorian paraffin lamps are fairly easy to restore and you can still get replacement wicks and shades, should you need them.

While most people prefer the idea of converting antique gas lights to electricity, it's not impossible to adapt early gas lights to run on modern gas supplies. The International Guild of Lamp Researchers (www.lampguild.org) is an American-based organization with a discussion board that always has lots of good suggestions for conversions and replacement parts, as well as a worldwide list of restorers. If you want to do more detailed research, The Historic Lighting Club in England is dedicated to all kinds of antique lighting, especially English lights, but covers all European lighting, including paraffin, gas, alcohol and other fuels. It should be stressed that the conversion, restoration and installation of any gas lamp should be carried out by a qualified person.

Chandeliers are a different matter. You can tackle cleaning an inexpensive chandelier yourself. First disconnect the chandelier from the electricity supply. Each glass droplet can be cleaned either in a solution of warm water with a touch of methylated spirit and white vinegar, or warm water with a splash of gentle washing-up liquid such as Ecover. Dry each glass droplet thoroughly and buff with a dry, soft cloth.

The cleaning and restoration of expensive chandeliers, however, is best left to the experts. Companies such as Woodall & Emery in West Sussex, Wilkinson's in London, and the Chandelier Group offer a variety of restoration services from cleaning and simple rewiring to total refurbishment. With an antique chandelier you may have to call upon a number of different disciplines – bronzing, gilding, gold leafing, wood carving, rewiring and wrought ironwork – so it's worth getting one company who can perform all these tasks rather than taking your chandelier to five different individual restorers.

USING SALVAGED LIGHTING

Before you buy any salvaged lighting, you'll need to decide where your new light fitting will go and what kind of effect you want it to produce. There are three main types of lighting – each important in its own right. Try to have at least one of each of these types of lighting in each room if possible:

GENERAL LIGHTING

This is also known as background lighting. It serves a very general function, providing illumination in place of daylight. It also provides a safety purpose – bad lighting can lead to accidents, especially on stairs and in bathrooms and kitchens. Used on its own, general lighting can be a bit dull, but team it with accent lighting and task lighting and you'll have a flexible, useful scheme. Examples of general lighting include ceiling-mounted lamps, chandeliers, wall lights, downlighters, uplighters and standard lamps.

ACCENT LIGHTING

This type of lighting is much more creative. It can be used to highlight a favourite painting or ornament, to create distinct areas of shadow and light, or even as a decoration in itself. You can create accent lighting with almost any type of light – spotlights, candles, downlighters, picture lights, sparkly fairy lights and table lamps, for example. Lampshades help to soften the effect of accent lighting, taking away the glare of the light bulb and helping to direct the light towards a specific area. If you want to highlight a specific object, you might want to consider the following:

- Spotlights and small strip lights can transform a bookcase or set of shelves by highlighting its contents.
- Light stained glass and glass ornaments from below or behind to enhance their translucent properties. If you have a glass vase, put fairy lights inside.
- It can be difficult to light a picture without getting glare from the glass. Use track spotlights or a picture light (which is mounted above the painting and pointed downwards). If the painting is particularly valuable you should check whether the heat and light from artificial lighting will cause any damage.
- Candles provide an intimate, shrine-like quality to a room. Metallic objects, gold leaf and other reflective surfaces such as mirrors work particularly well against candlelight.

TASK LIGHTING

This focuses light on a specific job you need to do, whether it's reading, sewing or cooking, for instance. It needs to be bright enough so you can see clearly, but directional so you aren't blinded by a harsh, radiating light. Examples include anglepoise lamps, clip-on spotlights and desk lamps. The best task lights are adjustable. A lamp or light can provide more than one type of lighting. A desk lamp, for example, can be used for task lighting but also makes effective accent lighting if turned to face a wall.

If you're looking for bargains, large green-enamelled steel lampshades and glass school globe shades are still good buys and look great in a contemporary kitchen. Defunct lights may also be used: decorative glass or ceramic lamp bases can easily be turned into candleholders. Simply strip the lamp base of all electrical parts and glue a wooden or metal candle plate to the top.

You can make a lamp out of salvaged items – a reclaimed jar, vase or champagne bottle, for example. If the object has no real monetary value you can drill a hole in the top to accommodate wiring and a lamp kit (available from most DIY stores). If you don't want to damage the object, however, you can get a socket cap that has a side outlet for the cord.

Old lighting fixtures can also be turned into different items. Glass globe shades make beautiful vases and tea-light holders (just make sure they aren't going to roll off your mantelpiece). Old glass lights also look great out in the garden: an inexpensive chandelier adds a real twinkle to a patio while vintage ceiling lights (the ones that look like bowls) make pretty hanging baskets.

EXPERT TOP TIPS

Anthony Reeve of LASSCO: Chandeliers and lights

- The vast majority of light fittings in yards and antique shops are twentieth-century.
- Signs of conversion from gas to electricity are reassuring, as they often indicate the light itself has some age to it.
- Look for makers' marks – often a small, applied plaque that might name a workshop such as 'Foster & Pullen of Bradford'.
- Look for old 'crown' glass with its inherent imperfections – replaced glass is immediately noticeable.
- The best quality chandeliers were hung with lead crystal, which has a clear bluish tinge. Most common is sodium crystal, which has a slightly yellow 'colour'.
- Get it checked – lights are usually sold for decorative value. Unless it comes with a certificate, second-hand lighting must be checked by a qualified electrician.

WINDOWS, STAINED GLASS AND SHUTTERS

Despite early examples of window glass, glazing was relatively rare until Tudor times. Right up until the fifteenth century, most domestic buildings just had *wind-holes* (from where the modern word 'windows' comes). These were simple, small openings designed to let out smoke and let in fresh air and light. To protect against bad weather, wind-holes would be covered up with wooden shutters at night. During the day, richer houses would cover their windows with translucent oiled parchment stretched across a trellis wooden frame. Poorer houses had to make do with distinctly less glamorous animal skin or cheap cloth.

At the beginning of the fifteenth century a few wealthier homes decided to incorporate glass into their windows. At first this simply meant filling in the existing wind-holes with small diamond or rectangular pieces of glass fixed together with lead strips (the technology for producing large flat pieces of glass was still some way off). These very early windows weren't designed to open and it wasn't uncommon to see only half the window glazed. A common shape in Renaissance windows, for example, was the four light cruciform: the top two lights were made from fixed leaded glass while the bottom two panes would have shutters or trellis frames.

CASEMENT WINDOWS

By the mid fifteenth century, glazing was becoming more common in grander houses and we also see the development of the casement window. Casement windows are side-hung and can be opened inwards or outwards depending on

how the frame is constructed. In general, casement windows open outwards in Britain and inwards on the Continent. (There was, however, an initial fashion in British towns for casement windows to open inwards to prevent passing traffic damaging the frame and its glass.) It's interesting to note, however, that window glass was still so expensive at this stage in the fifteenth century that you were well within your rights to take your windows with you should you decide to up sticks and move residence.

By the mid 1700s the fashion for casement windows had slowly spread from rich houses and urban centres to the humbler houses of the countryside, where they still remain the most popular type of window in traditional rural buildings. During the Arts and Crafts and Edwardian periods, mock-Tudor casement windows once again became fashionable in well-to-do houses.

SASH WINDOWS

During the seventeenth century, developments in glassmaking meant that larger pieces of window glass could now be manufactured, and in greater quantities. This not only changed the appearance and cost of casement windows, it also led to the development of a whole new style of window – the sash.

The sash window consists of two separate glazed frames which can be moved up and down to open and close the window. Opinions differ as to where it was invented, but the sash window seems to have been introduced from Holland or France in the middle of the 1600s, and it was an instant hit with architects. In fact, the name 'sash' comes from the French word *chassis*, meaning 'frame'.

At first the sash window sat flush with the front of the building and was kept open with a system of wooden pegs and notches. In 1709, however, the city of London decreed that, in the name of fire safety, all sash windows in the capital had to be recessed four inches into the wall. This move encouraged the development of the counter-weight system which allows sash windows to stay open without the need for pegs. The design for sash windows hasn't changed since.

In the early eighteenth century the horizontal sash window was introduced. Thought to be Yorkshire or Cornish in origin, these clever little windows didn't need a complicated counter-weight system because they opened by sliding sideways. Perfect for rooms with low ceilings, such as attics or cottages, the 'Yorkshire slider' became an equally suitable alternative to casement windows and is still a popular choice for rural renovation projects.

STAINED GLASS

In Britain, the earliest examples of stained glass date back to Saxon times, but it was the craftsmen of the Middle Ages who really developed stained glass into the art form that it is today. Window glass, as we know, was a rare and expensive commodity. Before the fifteenth century only cathedrals, monasteries and other wealthy religious establishments could really afford it. Stained glass not only offered a means of keeping out the elements but it also provided the perfect opportunity to pass on Christian teachings, using pictures instead of words, to the illiterate masses.

The secrets of stained glass window construction were described by the monk Theophilus in AD 1100 and the method has changed little since:

> If you want to assemble simple windows, first mark out the dimensions of their length and breadth on a wooden board, then draw scroll work or anything else that pleases you, and select colours that are to be put in. Cut the glass and fit the pieces together with the grozing iron. Enclose them with lead cames . . . and solder on both sides. Surround it with a wooden frame strengthened with nails and set it up in the place where you wish.

In other words, it's a bit like building a jigsaw puzzle.

After the Reformation the art of stained glassmaking declined and many of its precious skills were lost. Only later, in the nineteenth century, were the knowledge and techniques revived, helped by a resurging interest in medieval crafts and technology. Although early medieval stained glass is thought of as the finest, there are many exquisite nineteenth-century examples by companies such as Clayton & Bell and Morris & Co.

As well as intricately painted stained glass, both the Victorians and the Edwardians made good use of leaded glass windows. (Stained glass often refers to all types of leaded windows but the term should really only apply to glass that has been painted.) These are much simpler in construction, using clear pieces of uniformly coloured glass to create an attractive design. In an age obsessed with elaborate decoration, these coloured glass panels offered yet another opportunity to add visual interest to doors and windows all around the home. The largest collection of coloured glass would always be found in the entrance hall, deliberately placed there to impress visitors on their arrival and cast beautiful colours into the centre of the home. They also played a part in bathroom design, allowing light and privacy at the same time.

SHUTTERS

Shutters have been used to cover up window openings for centuries and even after the introduction of glass they still continued to be used as a means of blocking out light and keeping in heat (or, in hotter countries, keeping out heat). The Georgians were particularly keen on their shutters; early examples were made from solid planks of painted wood, but they weren't always practical: have them open and the room was in full sunlight (which might damage precious fabrics and paintings); have them closed and you were cast into instant darkness. The solution came in the form of the louvred shutter, a preferable model that would let in controllable amounts of light and air. Unlike European houses, shutters in Britain tended to be hung on the inside of the house, because it was impossible to close external shutters if you had outward-opening casement windows or low-opening sash windows. The exception to this rule is exposed coastal and rural houses, where external shutters provided extra protection from extreme weather.

In America and Europe external shutters are a common feature on houses even today. Their function is still important but, in this era of double glazing and curtains, many choose them for modern houses for aesthetic reasons more than anything else. It's difficult to imagine a Deep South plantation house or French chateau would look the same without them.

SOURCES

WINDOW FRAMES

If you're looking for an inexpensive way to replace an antique window, or plan to do an extension in keeping with an old property, carefully consider the advantages of salvaged windows. They're usually cheaper than new replacement windows and the old glass will look genuinely authentic. Salvaged windows, unless they're one-offs, usually come in sets, so you should be able to get enough to do the whole job. One thing to remember, however, is that they don't respond well to being adapted so it's better to buy exactly the right size or alter the opening to fit the window. Look out for salvaged window furniture too: you can pick up beautiful antique brass handles, hinges, locks and catches for just a few pounds each.

STAINED GLASS

In the 1950s and 1960s many Victorian and Edwardian homes had their leaded windows ripped out in the name of modernization. Many of these windows and door panels found their way to America, but a good selection has remained in the UK. Both these and newly salvaged leaded windows can be picked up at very reasonable prices at your local reclamation yard. Stained glass, on the other hand, is a much more precious commodity and fetches justifiably higher prices. Medieval stained glass is like the proverbial hen's teeth and for serious collectors and museums only. Nineteenth-century stained glass, however, is still available on the open market. Arts and Crafts pieces are highly collectable (for good reason – they're usually stunning) as are other nineteenth-century pieces. Keep an eye out for twentieth-century examples too – they can often be good quality and highly decorative. Unlike normal windows, stained glass is easy to adapt to a new opening and can be altered to fit.

If you're looking for stained glass of any quality, go through a specialist dealer; they will have the knowledge and expertise to help you choose, restore and fit a stained glass window into a new setting.

RESTORATION AND CLEANING

WINDOW FRAMES

Restoring a wooden window frame, whether casement or sash, shouldn't present too much of a problem for a skilled joiner; the design is basically the same as it always has been. Just make sure that they use the same quality of wood and mouldings, even if the window is designed to be painted. If you are buying a sash window, always replace the cords (from the counter-weight); if an old cord breaks once *in situ*, the results can be disastrous. The lower parts of sash windows are always the first to rot but it doesn't mean the whole window is useless; a skilled joiner will be able to cut out the rotten part and simply replace it with new wood (a process known as 'scarfing').

If the beading has dropped out, it'll need repairing – this is easily done with linseed putty and a putty knife. Simply squash putty into the edge of the pane with your thumb and forefinger, then pull the straight edge of the putty knife down the glass at a 45-degree angle to get a smooth finish.

LOUVRED SHUTTERS

If you want to restore an old painted pair of louvred shutters, bear in mind that stripping by hand will be very time-consuming. Don't use a blowtorch either – the fierce heat may damage the thin wooden slats and you'll end up driving yourself mad trying to get into all the complicated nooks and crannies. Many people choose to leave their old shutters with the romantic distressed look, a bit of shabby chateau chic, as it were; look for crackled paint finishes and shutters with layers of different-coloured paint showing through. If you can't bear to leave your shutters painted, the only real other option is to take them to be dipped in a bath of sodium hydroxide. Any repairs and repainting can then be easily carried out.

GLASS PANES

On older windows a common mistake is to replace broken panes with modern glass. Old glass, due to the manufacturing process, will often have slight imperfections, bubbles or rippling in the surface. This adds to the charm of the window and should, on no account, be viewed as a flaw to be removed. To get replacement glass for an old window, first ask your local salvage dealer – they may have a contact they use on a regular basis. Companies such as London Crown Glass can also supply old-style glass.

Genuine hand-blown cylinder glass is the best option if you want a truly authentic replacement windowpane. There are two types: low ream (LR) cylinder glass is best for sixteenth-century windows and leaded lights, no ream (NR) glass is suitable for windows of the late seventeenth century up to early twentieth century. Hand-blown glass is comparatively expensive, however, so if you're a bit strapped for cash, look for what's known as 'period style' window glass; this is cheaper than hand-blown glass but still looks much more authentic than perfectly flat modern float glass.

STAINED GLASS

To clean dirt from stained glass, use a soft cloth and a solution of warm water with a few drops of ammonia. Rinse clean with clear water and dry thoroughly with a dry soft cloth. The restoration of stained glass is, however, without doubt best left to the experts. A specialist dealer in stained glass will probably have a restorer on site, or at the very least, know who would be able to help you out. If you have no luck finding a restorer through a salvage dealer, contact the nearest church or cathedral with a good selection of stained glass: they will undoubtedly have consulted an expert stained glass restorer at some stage and will be only too happy to pass on their details.

USING WINDOW SALVAGE

Aside from using reclaimed windows and shutters for authentic restoration, you can find any number of fresh and creative uses for your salvage.

Shutters are perfect for creating a folding screen. Choose two or three of the same shape and style, and hinge them together. The taller thinner shutters make the most glamorous and useful screens, perfect for dividing up a room into different working areas or to use as a dressing area. They can also shield anything you don't want permanently on display – your computer, the TV, or the kitchen area if you live in a studio flat. Use them outside as a windbreak or as sun protection on a hot day.

Join four small shutters into a box shape and you have the base for a side table or stand. For an instant project, cover the back of a louvred shutter with a sheet of thin plywood. Hang the louvred shutter upside down on your wall as a letter rack for your home office or kitchen.

Wherever you would use a new window, consider a salvaged one, whether it's as a skylight, picture window, transom (window above a door), interior window or clerestory window.

Small window frames can be turned into mirrors or pinboards. Simply remove the existing panels of glass and replace with mirror or cork. You can also use a window as a picture frame: keep the old glass in place, arrange your photographs and then attach backing material to keep them in place. Old windows make fantastic trompe-l'oeil features too: just paint yourself a view and hang it on the wall.

Windows can be used as glazed cabinet doors or can turn a simple coffee table into a display table for curios. Decrepit old window frames can be used outside in the garden as frames for climbing plants such as sweet peas or ivy. They also make the basis for a very effective cold frame. Or what about using a glass-less window frame as a pot rack for the kitchen? Just suspend the frame from the ceiling by its four corners.

Leaded glass can be used in interesting, creative ways. Hang especially beautiful panels as if they were paintings; to get their full effect make sure they are suspended in front of a window or a light. Large leaded windows set into a wooden frame also make the most fantastic room dividers. Small pieces of stained glass make lovely ornaments, propped up on a window ledge or in front of a table lamp so the light can shine through.

EXPERT TOP TIPS

Drew Pritchard: Stained-glass windows

- Always deal with a professional, someone who has years of hands-on experience and is workshop trained. Ask to see some of their work.
- Most stained glass windows can be altered to fit any space – take advice!
- Door panels and large windows need saddle bars (horizontal steel bars lined up on the back of windows) to support them.
- The value of religious stained glass and painted glass is dictated by the design and quality of painting, the glass colour choice and the condition of leadwork. Also, some makers, painters and designers can command a premium.
- Pay particular attention to the hands and faces of any figures.
- Never break up complete sets; if they were designed together they should stay together.
- Don't be put off by cracks: originality is always my prime concern when buying a window. Above all, buy what you love – after all, you will be looking at it for a long time!
- Leave cleaning, repairs and removal to a professional.

WOODWORK AND STAIRS

From the earliest civilizations, man has used timber in his home. Abundant and strong, wood has been exploited as a basic building material as well as providing the perfect base for decorative carving. The variety of wooden architectural items is almost limitless, but salvage yards are a particularly good source of staircases, panelling and structural components such as beams. (See Floors and Roofs, page 106, for information on reclaimed floorboards.)

Most early cottages and houses didn't bother with a staircase – in fact, until recently you could still find examples in Britain of small vernacular buildings which used a ladder to get up to the first floor. Important monastic buildings or wealthy houses would have had stone staircases, but it wasn't until the second half of the seventeenth century that timber staircases – usually made from oak – became popular. Since that time the basic design has remained the same.

Each staircase has a set of standard components. The stairs are made up of three parts: the horizontal tread, the nosing (overlapping lip) and the vertical rise. The railings, or balustrade as they are properly known, consist of balusters (individual, often carved spindles) held together by the handrail. At the bottom and top of the stairs you have large, chunky newel posts keeping the whole staircase firmly in place.

Until the Victorian age staircases were hand-crafted, high-status investments. One of the most expensive architectural items in the home, the staircase was designed to reflect the wealth and good taste of its owner. Ornamentation soon become a key factor in the design of any staircase: newel posts were decorated with intricate designs and topped with fancy finials, while balusters were carved into ever more complex shapes such as barley twists and tapered columns.

With the introduction of industrialization in the eighteenth and nineteenth centuries, wooden staircases became simpler to make and cheaper to produce. Middle-class homes could now afford the extravagant staircase designs previously only available to the super-rich, and even relatively modest homes still had staircases with a little flourish. These remained popular well into the early twentieth century and are a key feature of many Victorian and Edwardian homes. By the 1930s, however, the new trend for simple, utilitarian designs supplanted the traditional, showy staircase of the past.

Wood panelling has been used in wealthy homes since medieval times; it served a decorative purpose and provided a welcome extra layer of insulation against the cold. Both polished hardwood and painted softwood panelling was used in grand houses in the fifteenth century and carried on being popular right up until the beginning of the eighteenth century, when it seemed to fall out of favour.

A hundred years later, the Victorians revived panelling with great enthusiasm, using dark hardwoods to create a rich background for their highly ornate furniture and fabrics. The fashion for panelling continued well into the twentieth century with the Edwardian fashion for simple, stylish panelling that often reached two-thirds of the way up the wall and stopped at a narrow shelf. Even as late as the 1930s oak panelling was a common feature of surprisingly modest homes: mock-Tudor in design, the panelling took its influence from the Arts and Crafts movement and matched the obligatory oaks doors and false beams of the time.

Unlike panelling, skirting board is rarely rescued from a building site as it can be difficult to remove and often gets damaged in the process. This is a pity, however, as skirting differs greatly throughout the historical periods and yet only a few styles of modern skirting are available to the prospective renovator. Early eighteenth-century skirting boards, for example, were usually very low to the floor whereas Greek Revival houses of the late eighteenth century had much taller, grander skirting boards – some were even made from marble. As a general rule, however, tall rooms require tall skirting boards.

Timber has been used as a structural building material from time immemorial – it has excellent compression, tension and bending strength, and in spite of appearances it's actually relatively fire resistant. Thick beams, especially if they are made from hardwoods such as oak, burn slowly, forming a protective layer of charcoal on the outside.

In Britain, oak and elm were commonly used for structural components in all manner of buildings, from poor housing to palaces. Medieval timber-framed houses ranged from simple post and beam structures to elaborate A-framed and box-framed buildings with galleries and upper storeys. Timbers of the day were joined together using mortise and tenon or dovetail joints anchored with thick wooden pegs – not a metal fastener in sight.

As hardwood timber supplies began to dwindle during the seventeenth century, and disasters such as the Great Fire of London of 1666 shook people's faith in timber buildings, brick and stone became popular replacement building materials. For the next two hundred years timber was still freely used in the construction of buildings – especially in roofs, ceilings and floors – but it didn't dictate the interior and exterior design of the building as it had once done. Great pains were often taken to disguise any structural timber in Georgian and Victorian houses, for example, and you'll only find exposed beams in attics, traditionally the servant's quarters.

SOURCES

As with all salvage, you'll find a good selection of timber at a general salvage yard and a huge supply at a specialist dealer. Demolition yards and timber merchants will also have reclaimed material in stock, although if you are looking for wood products of any real antiquity or interest, architectural salvage yards are probably your best bet.

Why use salvaged timber? There's a strong ecological argument for reusing wood. In the past, timber was cut from the world's precious old-growth forests – trees that take hundreds of years to regrow. Reusing old timber not only makes the most of this precious resource but it also reduces the pressure to fell additional old-growth forests.

Old-growth timber, because it matures so slowly, is aesthetically more pleasing than new fast-grown wood. Slow-growth timber is dense and heavy. It has a fine, straight, tight grain and is often free from knots. Antique timber has had time to season fully, so it shouldn't warp or shrink: this makes it the ideal material for floorboards, beams and structural components. This natural seasoning process will also have brought out the subtle colour differences which are unique to each tree species.

Fans of reclaimed timber will tell you that, hand in hand with its superior ecological status and aesthetics, there's no comparison in terms of craftsmanship. The production and processing of timber has changed dramatically since the early days of hand-carved beams and floorboards. Timber is now produced

industrially at huge speeds and each board or beam is uniform in shape, texture and design. This is handy for the building trade, but something has been lost in the process. The character and hand-made quality of hand-hewn or pit-sawn timber has an integrity and appeal that simply can't be matched by industrially processed wood.

Costs vary greatly, depending on what you want – a reclaimed railway sleeper can be picked up for next to nothing while antique panelling is always expensive. What you pay will reflect the amount of effort it took to salvage the timber in the first place. A typical lumber (building timber) salvage project will involve a wooden structure being carefully dismantled and then laboriously cleaned of all visible metals such as nails, screws or brackets. If the timber is dirty, it might also need pressure-washing. The timber will then need to be sorted in terms of sizing and graded, ready to be sold. Decorative timber will be treated differently in a salvage yard: on no account should anyone be pressure-washing antique panelling or fretwork. These types of woodwork may need careful restoring, however, and this needs to be factored into the cost.

There's a fantastic variety of salvaged woodwork up for grabs. Alongside stairs, panelling and beams, other reclaimed timber includes flooring, skirting board, doors, window frames, shutters, friezes, fretwork and decorative carving such as bosses. Look for old timber from factories and workshops: barrel makers (cooperages), for example, produce some wonderful pieces. In America the prevalence and scale of barn buildings means there's always a wonderful supply of antique crafted wood and beams for any potential builder or renovator.

Environmental groups are thrilled at the increasing popularity of using all these different sorts of reclaimed timber; according to Friends of the Earth, over three thousand tonnes of reusable wood is thrown away from the demolition of old buildings in the UK every working day. This is such a shame, because timber is the easiest material to re-work and re-shape. The very useful Friends of the Earth *Good Wood Guide* has lots of top tips for buying recycled and reclaimed timber, including:

- Check what sizes of timber are easily available from salvage yards before designing DIY projects (this especially applies to doors, windows and staircases).
- Get the best timber from older buildings.
- Thoroughly inspect your timber salvage and check every detail. Take your builder with you if you are unsure. Ask to look inside packaging – reputable dealers won't mind.

- Whatever timber salvage you choose, you'll also need to ensure that it's free from woodworm, dry rot and other potentially harmful problems. Timber hates damp, so make certain that your salvage has been stored in a dry location; wet wood will not only be subject to rot, but it will also split and warp when introduced to a warm, dry house.

Active woodworm needs to be treated, but evidence of long-extinct woodworm needn't be a cause for concern. Timber of any real age will be lucky to have escaped some form of attack and the delicate patterns left behind can be quite a feature (not everyone shares this enthusiasm for woodworm tracery, however, so be careful if you are hoping to attract potential buyers). It's also worth noting that serious woodworm activity can affect the structural integrity of timber: if you need a piece of wood to be load bearing, make sure that it is up to the job. If you're not sure, get a structural engineer's advice.

Dry rot is a disaster if you bring it inside the house. It thrives in damp conditions and loves badly ventilated areas. Look out for the characteristic cotton wool-like fungal spores and a strong smell. Wet rot, on the other hand, has black cobwebby fungal strands and will appear on wood that is black or dark brown. Wet rot usually only affects the damp part of a piece of timber, whereas dry rot can run amok and start damaging surrounding building material.

To spare yourself any future problems, look for wood that is sound, dry and free from any evidence of fungal spores. Softwoods are more susceptible to infection than hardwoods. Hardwoods aren't totally immune to disease, however, and on no account should you buy a lightweight piece of oak as there is likely to be something seriously wrong with it. When you are buying reclaimed beams, a top tip is to look for a beam that has had its end cut off with a chainsaw: the newly exposed wood will give you a good idea of the internal state of the rest of the beam. Get any old wood treated before you bring it into the house. If you don't want to use toxic sprays, there are eco-friendly options available such as borax. It's also worth noting that woodworm thrives in damp wood, so the best defence is to keep your timber dry.

If you are planning to use a large amount of reclaimed timber in a building project, leave yourself lots of time to source the right materials. The timber salvage yard may not have exactly what you want in stock but might be able to source it from elsewhere – another supplier or a forthcoming demolition project. This can take some time. Always establish how long it will take for your timber to arrive, so you can factor the lead time into the design and construction of your building.

RESTORATION AND CLEANING

Because of the wide variety of timber salvage available, it's difficult to give general advice about restoration and cleaning. There's a big difference between pressure-washing a rustic beam and carefully restoring the finish on Georgian panelling, for example. On the whole, less is more with salvaged timber. Over-zealous and inappropriate cleaning and stripping might cause irreparable damage and remove years of valuable patina. Equally, reapplying the wrong finish can do your timber no favours; in general, stick to old-fashioned methods of wood finishing if you want an authentic, sympathetic look: in other words, oils, waxes and flat eggshell paints rather than polyurethane varnishes and modern high-gloss paints. If you are spending serious money on reclaimed woodwork – especially expensive panelling, carving or other decorative pieces – consult an experienced restorer who will be able to tell you at a glance what the right course of action should be.

REVIVER

There is, however, one quick way to enhance any polished woodwork you may acquire. Professional restorers use what is called 'reviver' in the trade: it enhances the grain and re-establishes shine to finishes dulled with grease and grime. Recipes vary, but you can use one part raw linseed oil and one part white spirit, with a dash of vinegar. The white spirit will gently remove the top layer of polish, the vinegar cuts through any grease and the linseed oil lubricates and adds shine. Simply rub on with a soft lint-free cloth, using circular motions. Rub off again quickly with another soft cloth. (N.B. Linseed oil will darken wood if it is allowed to sink in.)

STRIPPING AND DIPPING

As discussed in the Doors and Door Furniture chapter, the current fashion for stripping back every piece of wood to its natural finish isn't necessarily histori-cally accurate. Take panelling, for example. Most softwood panelling of the past was designed to be painted, whereas hardwood panelling (which was more expensive) may have been left bare. If you do leave your panelling in its natural state, don't use a normal varnish on it – use oil or wax polish for an authentic, softer finish. There's one exception, however: wooden shutters may have to be dipped as they are notoriously fiddly to strip by hand.

SANDING

Never plane or sand down antique woodwork unless absolutely necessary: the old patina will disappear, along with most of the timber's appeal. Old timber isn't supposed to be free from marks, dents and scratches. In fact, timber with original hewn markings, mortise or bolt holes, maker's stamps or carvings and other evidence of work will often have greater monetary and aesthetic value than plain unmarked wood.

USING WOODWORK SALVAGE

In general, reclaimed timber has a unique colour and texture that you simply can't get from new wood. Reclaimed flooring, for example, looks superb in older buildings and complements antique furniture, but works well in contemporary, urban designs too. Make the most of wear and tear: a lovely old smoothed wooden bench is much preferable to a brand-new, characterless seat. Wear such as this is impossible to reproduce artificially and will add value to a piece.

Beams can be used for their original purpose, as ceiling rafters or support pillars, but consider using them for other functions too – as fireplace or doorway lintels, for example. (N.B. Oak is by far the safest wood to use for fireplaces.) They also make chunky shelves. Consider using beams to make a dramatic four-poster bed, a porch or a gazebo in the garden. Cut them down into smaller sections and turn them into table tops, work counters or rustic butcher's blocks.

Railway sleepers are commonly employed in the garden. Used as steps, decking and other landscaping features, they're often a cheap and attractive way to bring structure to outside spaces. It's worth bearing in mind, however, that almost all railway sleepers will have been treated with tar and creosote, which has a tendency to seep out in warm weather. This isn't a problem unless you are planning to use the sleepers for kids' play equipment or where they'll come into contact with bare skin (i.e. benches or outdoor tables). If you're lucky, you can find untreated railway sleepers; these tend to be made from hardwood and will cost more. You can also sometimes find old salt-treated softwood sleepers, but again these are fairly unusual.

Large sections of panelling can, of course, be used for their original purpose, but if you find yourself with smaller pieces consider using them as cupboard doors, screens or bed boards. If you are lucky enough to find a particularly decorative piece of panelling, consider mounting it as a picture.

Old weathered wood looks wonderful in an interior. It comes in all different colours – from rich dark brown to light bluey grey – and often has a grooved, deeply tactile texture. Designers love its distressed, well-worn patina and it makes wonderful picture and mirror frames. Old weathered shutters also look gorgeous: resist the temptation to repaint them and, instead, enjoy their faded appeal.

STAIRCASES

Salvaged staircases aren't the easiest objects to install. Each staircase was built to fit a particular house design (sometimes even a particular house) so it's not just a question of plonking your reclaimed staircase into your hallway. In fact, it's easier to build your house around your staircase. Either that, or choose the staircase very early on in the restoration of a house, so you can get a builder to alter the hallway accordingly.

If you have to replace your staircase and you don't want to make structural changes to your home, consider employing a specialist joiner to make a copy. This may be expensive but the result will be far more pleasing than attempting to fit a modern mass-produced staircase, the dimensions of which will probably be totally wrong for an old house.

If you want to include an old salvaged staircase in your home, you'll need to consider modern laws. Building regulations have a lot to say about headroom, handrails, the pitch of the staircases and so on. Old staircases are often too steep, too narrow or not deep enough for modern regulations, which is a pity, as many are incredibly beautiful and eminently safe. Talk to your local planning officer: if you can prove that your salvaged staircase is replacing like with like (in other words, you are simply putting back in the original-style staircase) your plans may be exempt from the normal rules. Putting an old staircase in a new building or extension, however, may not.

If you are attempting to restore an existing staircase with pieces salvaged from a yard, remember to match the right period and style. A Regency hand-made

staircase, for example, will be ruined with Victorian machine-turned balusters. One good point, however, is that staircase design hasn't really changed over the past three centuries so a skilled joiner should know how to tackle most repairs.

Of course, you don't have to use your salvaged staircase for its original purpose. The spindles are endlessly useful and make wonderful supports for tables, birdfeeders and benches. Use them individually as dramatic candlesticks and lamp bases, or sideways as towel rails. Use them in groups as radiator covers or wherever you need a set of railings. Mount spindles on wooden blocks and turn them into a lawn chess set.

Newel posts can also be used in isolation: anywhere that needs a sturdy column will be well served by a salvaged newel post. Put four together and you've got four corners of a wooden bed. Very dramatic newel posts and finials are often beautiful enough to have as pieces of carving in their own right.

EXPERT TOP TIPS

Alan Bitterman of Rarefinder: Woodwork and stairs

- When it comes to buying reclaimed staircases, it's all about perspective – make sure the stairs are right for your house.
- The biggest mistake you can make is not employing an experienced joiner who knows how to work with reclaimed materials.
- Never strip reclaimed wood, and never varnish it.
- Using reclaimed wood is never easy – be prepared to spend money to make it look right.
- You must be happy with the patina on woodwork – that's what makes it so special.
- Look for loose sawdust around woodwork – it means that woodworm is active and you will have to go through the whole process of removing it.

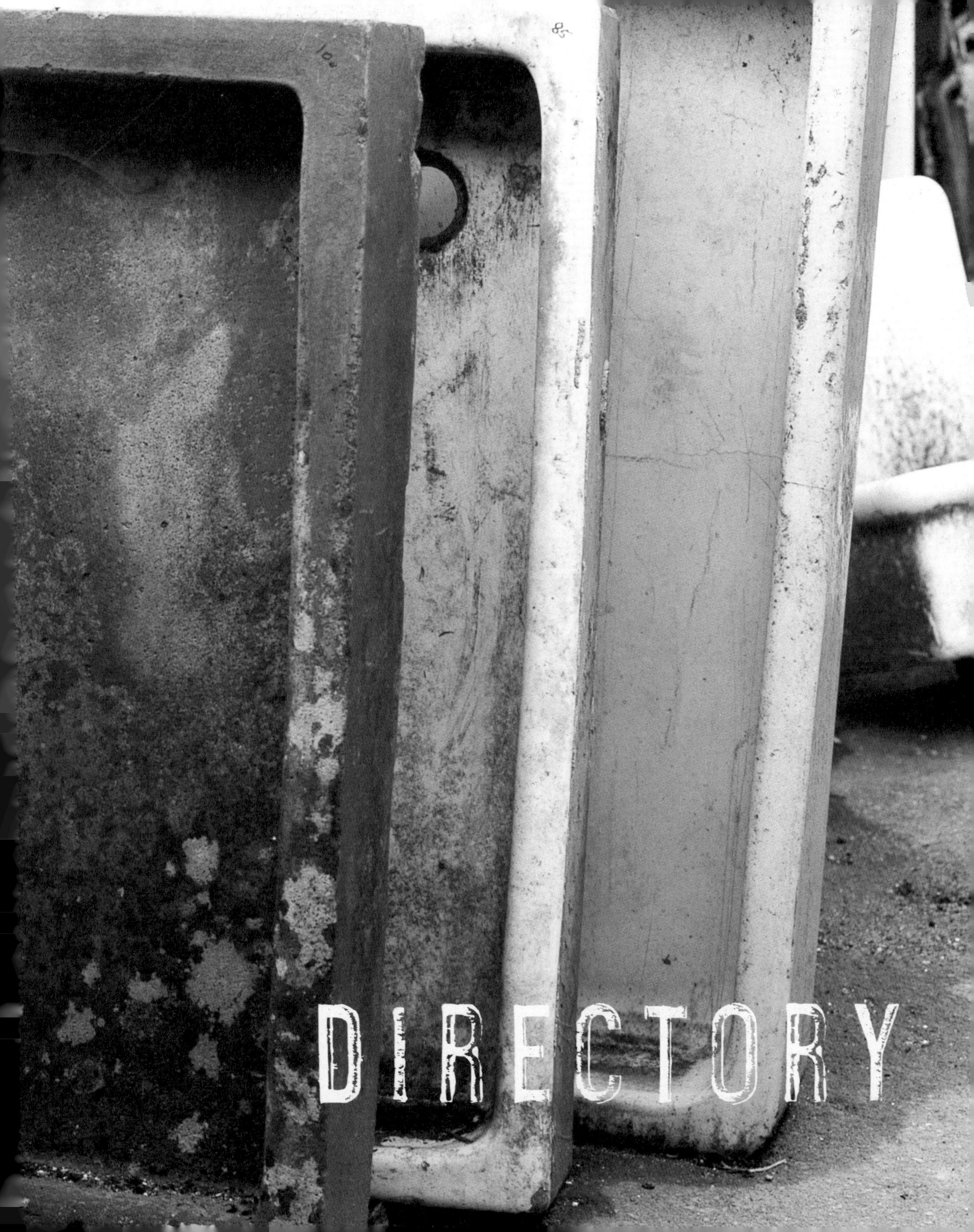

DIRECTORY

THE SALVO CODE

Since 1990 Salvo has been involved in the establishment of a simple code for dealers who buy and sell architectural antiques, antique garden ornaments and reclaimed building materials.

Salvo's aim has been to give buyers confidence that items they buy have not been stolen or removed from listed or protected buildings without permission.

Many dealers have already established a sensible buying procedure but the Salvo Code makes this more formal, understandable and obvious to the buying public.

In this way customers are given the choice of buying from relatively safe and responsible sources.

Each Salvo Code dealer has a Certificate of the Salvo Code, which is dated for the current year and is signed by Thornton Kay, the administrator of the Salvo Code.

THE CODE

The Salvo Code Dealer undertakes:

1. Not to buy any item if there is the slightest suspicion that it may be stolen.
2. Not to buy knowingly any item removed from listed or protected historical buildings or from sites of scheduled monuments without the appropriate legal consent.
3. To record the registration numbers of vehicles belonging to persons unknown to it who offer items for sale, and to ask for proof of identity.
4. Where possible to keep a record of the provenance of an item, including the date of manufacture, from where it was removed, and any previous owners.
5. To the best of its ability and knowledge, to sell materials free from toxic chemicals, excepting those natural to the material, traditional to its historical use, or resulting from atmospheric pollution.
6. To allow its business details to be held on a list of businesses who subscribe to the Salvo Code and to display a copy of the code and their Certificate in a public position within their business premises.

USEFUL ADDRESSES

And now for the important bit. Below you'll find a selection of the various salvage yards, restorers, craftsmen and related suppliers in your area. Salvage dealers who have signed up to the Salvo Code are indicated with the Crane logo. At the end of the Directory you'll also find a number of national organizations who can help you find out more about period details and restoration.

ENGLAND

BEDFORDSHIRE

Architectural Antiques
70 Pembroke Street
Bedford
Bedfordshire MK40 3RQ
01234 213131
Architectural antiques. Period chimneypieces, sanitary ware, doors, tiles, radiators, furniture and garden statuary. Also specialize in supply and installation of period fireplaces in marble, stone and cast iron as well as hardwoods and softwoods.

The Bath Works
Glenacres
Watling St.
Dunstable
Bedfordshire LU6 3QS
01582 602713
www.thebathworks.com
Antique bathrooms. Renovation and supply of antique baths. Also have a shop in the USA selling antique baths.

Capital Fireplaces Ltd.
Unit 12-17 Henlow Trading Estate
Henlow Camp
Bedfordshire SG16 6DS
01462 813138
www.capitalfireplaces.co.uk
Iron and steel. Replacement spare parts for cast-iron fireplaces. Ash covers, frets, bars, hoods, etc.

K&C Roofing
East Siding
Sandy Station
Sandy
Bedfordshire SG19 1AW
01767 680213
kcsandy@kbnet.net
Roof slates and tiles. Roofing contractors and reclaimed and new roofing merchants.

Nationwide Reclaim
Goosey Lodge Farm
Goosey Lodge Industrial Estate (off A6)
Wymington Lane
Wymington
Bedfordshire NN10 9LU
01933 313121
info@nationwidereclaim.co.uk
www.nationwidereclaim.co.uk
Timber and stone. Specializing in reclaiming quality timber and quarry

materials and preparing for refurbishment or new-build projects.

Olde English Reclamation
2b Conquest Mill
off Station Road
Ampthill Industrial Estate
Ampthill
Bedfordshire MK45 2QY
01525 406662
oer@btconnect.com
Reclamation yard.
Bricks, roof tiles and slates, doors, beams, flooring, fireplaces, bathrooms, radiators and pews.

BERKSHIRE

Jonathan H. McCreery
Lambourn
Berkshire
01488 71367
Antique gardenware.
Also Victorian hall and umbrella stands.

BRISTOL

Au Temps Perdu
28-30 Midland Road
St Philips
Bristol BS2 0JY
01179 299143
www.autempsperdu.com
Architectural antiques.
Reclaimed building materials and restoration.

Chauncey's
15-16 Feeder Road
St Philips
Bristol BS2 0SB
01179 713131
sales@chauncey.co.uk
www.chauncey.co.uk
Reclaimed flooring.
Traditional floorboards, woodblock, woodstrip and new hardwood floors.

Greyfield Timber
4 Maynard Terrace
Clutton
Bristol BS39 5PL
01761 453760
Timber. Timber reclaimer, kitchens from reclaimed wood.

Martyn Lintern Carpentry
Bristol
01179 660294
07946 519441
Carpenter and joiner.
Hand-built furniture and kitchens from reclaimed materials.

Robert Mills Ltd.
Narroways Road
Eastville
Bristol BS2 9XB
01179 556542
www.rmills.co.uk
Architectural antiques.
Antiques for pubs and restaurants. Panelling, stained glass, church fittings, reredos, pulpits, prie-dieux and other religious artefacts.

Olliff's Architectural Antiques
21 Lower Redland Road
Redland
Bristol BS6 6JB
01179 239232
www.olliffs.com
Architectural antiques.
Sourcing clients' needs. Trader in all types of antique architectural components, garden ornaments, gazebos and summerhouses.

Rose Green Tiles & Reclamation
206 Rose Green Road
Fishponds Trading Estate
Fishponds
Bristol BS5 7UP
01179 520109
www.rosegreenreclamation.co.uk
Reclamation yard.
Reclaimed roof tiles and slates, ridge tiles, bricks, cast-iron radiators, new oak railway sleepers, flagstones, floorboards, oak barrels, chimney pots, troughs, staddle stones and garden ornaments.

BUCKINGHAMSHIRE

Dismantle and Deal Direct
108 London Road
Aston Clinton
Buckinghamshire HP22 5HS
01296 632300
www.ddd-uk.com
info@ddd-uk.com
Architectural antiques.
Doors, windows and fire-surrounds, as well as garden antiques and statuary.

Rory Duncan
Beech Cottage
Nether Winchendon
Aylesbury
Buckinghamshire HP18 0DY
01844 299529
Architects and designers.
Reclamation-friendly architect.

IBS Reclaim Ltd.
Thame Road
Oakley
Buckinghamshire HP18 9QQ
01844 239400
Reclamation yard. Reclaimed York stone, setts, bricks, tiles, slates, doors, flooring, fireplaces and statuary.

Just Slate Ltd.
Wayside Farm
Fleet Marston
Aylesbury
Buckinghamshire HP18 OPZ
01296 655131
Roof slates and tiles. Reclaimed slates, tiles and bricks.

Midland Stone Co. Ltd.
High Street
Lavendon
Buckinghamshire MK46 4HA
01234 712712
Reclamation yard. Mostly reclaimed stone, but also bricks, tiles and other reclaimed building materials.

Site 77
College Road North Business Park
Aston Clinton
Buckinghamshire HP22 5EZ
01296 631717
www.site77.com
Reclamation yard. Reclaimed bricks, tiles, slates, oak, doors, York stone, radiators, quarry tiles, setts and flooring.

CAMBRIDGESHIRE

Emmaus Cambridge
Green End
Landbeach
Cambridge
Cambridgeshire CB4 8ED
01223 863657
www.emmaus.org.uk
Second-hand shop. Second-hand furniture, books, electricals, records, clothes, kitchenware, bric-a-brac.

Snailriver Co.
Fordham
Ely
Cambridgeshire
www.snailriver.co.uk
Leadworker. Antique and reproduction garden ornaments for sale, restoration and bespoke leadwork, mould and pattern makers.

Solopark plc
Station Road
near Pampisford
Cambridgeshire CB2 4HB
01223 834663
www.solopark.co.uk
Reclamation yard. Major suppliers of reclaimed building materials, bricks, clay roof tiles, period architectural antiques, garden antiques and reproductions; dismantling, reclamation and demolition contractors.

Window Doctor
The Grange
Biggs Road
Wisbech
Cambridgeshire PE14 7BE
01945 584885
Glassworker. Restores windows and doors, glass and locks.

CHESHIRE

Artisan Genuine Stained Glass
21 Tatton Road
Sale
Cheshire M33 7EB
0161 962 8090
www.genuinestainedglass.com
Glass. Stained glass artist.

Ashley Reclamation
Brickhill Yard
Brickhill Lane
Ashley
near Altrincham
Cheshire WA15 0QF
0161 941 6666
www.ashleyreclamation.co.uk
Reclamation yard. Reclaimed hard-landscaping materials, ground coverings and floors, brickwork and other stone, doors, windows, radiators and stairs.

Beeston Reclamation
The Old Coal Yard
Beeston
Tarporley
Cheshire CW6 9NW
01829 260299
Reclamation yard. Reclaimed building materials, architectural antiques, dismantling and demolition.

Cheshire Demolition & Excavation Ltd.
Moss House
rear of 72 Moss Lane
Macclesfield
Cheshire SK11 7TT
01625 424433
www.cheshiredemolition.co.uk
Reclamation yard. Reclaimed stone, bricks, slates, tiles, oak and pine beams, doors, fireplaces, gateposts, coping and architectural items.

Core Design (UK) Limited
26 Shaw Road
Heaton Moor
Stockport
Cheshire SK4 4AE
0161 443 4000
www.core-design.net
Architects.

English Garden Antiques
The White Cottage
Church Brow
Bowdon
Cheshire WA14 2SF
0161 928 0854
www.english-garden-antiques.co.uk
Antique gardenware. Garden antiques, ornaments, sundials, troughs, planters, urns, millstones, staddle stones, pumps, farm tools.

Grosvenor Stone
Golborne Bridge Farm
Handley
Cheshire CH3 9DR
01829 770632
www.grosvenorstone.co.uk
Composition stone. Manufacturer and supplier of natural and reconstituted sandstone products, repro architectural dressings and garden statuary.

Lockside Estates Ltd.
Lockside Mill
St Martins Road
Marple
Stockport
Cheshire SK6 7BZ
0161 427 7723
Architects. Conservation architects and historic building consultants.

Nostalgia
Hollands Mill
Shaw Heath
Stockport
Cheshire SK3 8BH
0161 477 7706
www.nostalgiafireplaces.co.uk
Antique fireplaces. Over 1,400 reclaimed fireplaces of all descriptions, marble chimneypieces a speciality. 200-plus pieces of antique sanitary ware.

R&R Renovations & Reclamation
Canalside Yard
Audlem
Cheshire CW3 0DY
01270 811310
Reclamation yard. Reclaimed bricks, stone, oak beams, tiles, slates, oak barn frames, architectural antiques, bathrooms; dismantling contractors; brick buildings sought.

John Shone
7 Sandringham Road
Widnes
Cheshire WA8 9HD
01514 249189
www.antiquegrandfatherclocks.com
Antique furniture. Specialist in the buying, selling and renovation of fine antique grandfather clocks ('longcase' or 'tallcase' clocks) for over thirty years.

SOS Ltd.
Hawkswood
Coach Road
Little Budworth
Cheshire CW6 9EJ
01829 760085
Reclamation yard. Reclaimed materials and large stock of beams; also new hardwood and softwood flooring.

CUMBRIA

Carlisle Demolition
Carleton Depot
London Road
Carlisle
Cumbria CA1 3DS
01228 530099
Demolition contractors. Reclaimed materials and scrap metal.

Mike Gibstone
5 Scawfell Crescent
Seascale
Cumbria CA20 1LF
01946 727122
Mason. Stonemasonry, fixing and repair service specializing in reclaiming the past.

Lakeland Timber Reclamation Co.
9 Kendal Green
Kendal
Cumbria LA9 5PN
01539 730097
www.ltrco.com
Timber. Reclaimed timber, oak and pine beams, woodstrip, wood block, wide oak flooring, pine floorboards, parquet de Versailles panels made in the authentic style with reclaimed old oak, and reclaimed flagstones, mainly York stone.

Richard Leask
The Corn Mill
Warwick Bridge
Carlisle
Cumbria CA4 8RE
01228 562430
Reclamation yard. Reclaimed flooring, flags, fireplaces, beams, stained glass, doors.

Wilson Reclamation Services Ltd.
Yew Tree Barn
High Newton
Grange-over-Sands
Cumbria LA11 6JP
01539 531498
www.yewtreebarn.co.uk
Architectural antiques. Architectural antiques and salvage, garden artefacts, fireplaces, antique furniture and tea rooms (on the A590).

DERBYSHIRE

Glens Reclamation Ltd.
220 Max Road
Chaddesden
Derby
Derbyshire DE21 4HB
01332 386037
Reclamation yard.
Roofing slates and tiles, chimney pots, oak and pine beams, pavers, oak flooring. Open by appointment only.

Landmark Landscapes
Netherside
Hope Valley
Derbyshire S33 9JL
01433 623132
Reclamation yard. Reclaimed York stone, paving, troughs, pine flooring, lampposts, radiators and garden benches.

Sheffield Architectural Salvage
Moor Grange Farm
Moor Lane
Taddington
Buxton
Derbyshire SK17 9RA
01298 85020
Architectural antiques.
Cast-iron radiator specialists.

Smith Hey Stone
Hilton
Derbyshire
01283 734161
07860 467877
www.smith-hey-stone.co.uk
Antique gardenware. Antique stone troughs, staddle stones, terracotta and cast-iron garden ornaments, dressed and unusual stone.

DEVON

Antique Baths of Ivybridge
Erme Bridge Works
Erme Road
Ivybridge
Devon PL21 9DE
01752 698250
www.antiquebaths.com
Antique bathrooms.
Antique bathroom equipment bought and sold. Full bath re-enamelling service, *in situ* or in workshop.

Ardosia Slate Co. Ltd.
16 Park Street
Lynton
Devon EX35 6BY
01598 753642
Slate. Importers of stone and slate; flooring, sills, tables, worktops.

EW Trading Ltd.
18 Mill Street
South Molton
Devon EX36 4AR
01769 574147
Reclaimed flooring.
Reclaimed oak flooring and terracotta floor tiles, and new, imported from France. Warehouse in Barnstaple.

Kenmart Timber
Lower Mill
Pottery Road
Bovey Tracey
Devon TQ13 9JJ
01626 833564
www.timberandslate.co.uk
Reclamation yard. Reclaimed and new slates, tiles, ridge tiles, chimney pots, granite setts and lintels, flooring, garden items.

Stax Reclamation
Agaton Farm
Budshead Road
Ernesettle
Plymouth
Devon PL5 2QY
01752 363838
www.staxreclamation.co.uk
Reclamation yard.
Reclaimed building materials and architectural salvage.

Tobys Architectural Antiques Ltd.
Station House
Station Road
Exminster
Exeter
Devon EX6 8DZ
01392 833499

Brunel Road Industrial Estate
Newton Abbot
Devon TQ12 4PB
01626 351767

Torre Station
Newton Road
Torquay
Devon TQ2 5DD
01803 212222
www.tobysreclamation.co.uk
Architectural antiques.
Three locations, all offering architectural antiques, garden statuary, unusual gifts, pine, hardwood, slate and stone flooring, sanitary ware, original fireplaces, windows and doors.

Graham White
South Molton
North Devon
07887 593441
www.grawhite.com
Reclamation yard. Supplier of antique stone for home and garden, period fireplaces and salvage.

Winkleigh Timber
Seckington Cross Industrial Estate
Winkleigh
Devon EX19 8DQ
01837 83573
Flooring. Large stock of reclaimed timber, including flooring, beams, furniture. Also slate floors.

DORSET

Ace Reclamation
Pine View
Barrack Road
West Parley
Wimborne
Dorset BH22 8UB
01202 579222
www.acedemo.co.uk
Reclamation yard. Reclaimed materials, timber, bricks, tiles, RSJs, slates, architectural items, recycled concrete, sleepers, telegraph poles; demolition and dismantling contractors.

Dorset Reclamation
Cow Drove
Bere Regis
Wareham
Dorset BH20 7JZ
01929 472200
www.dorsetreclamation.co.uk
Reclamation yard. Reclaimed traditional building materials, architectural antiques, garden ornaments, bathrooms and fireplaces.

Gardentique
Lower Common Lane
Three Legged Cross
Wimborne
Dorset BH21 6RX
01202 829333
www.gardentique.co.uk
Antique gardenware. Antique and new classical garden ornament, furniture and statuary.

Minter Reclamation
Lower Common Lane
Three Legged Cross
Wimborne
Dorset BH21 6RX
01202 828873
www.minterreclamation.co.uk
Reclamation yard. Reclaimed building materials and architectural antiques.

Oldays Ltd.
64-70 Bridge Street
Christchurch
Dorset BH23 1EB
01202 488980
www.oldays.co.uk
Reclamation yard. Reclaimed building materials, stone, old and new flagstones, roof tiles, old beams, architectural and garden antiques.

Purbeck Engineering
Knoll House
Studland Bay
Swanage
Dorset BH19 3AH
01929 450450
Ironwork restorers. Occasional old ironwork for sale.

Semley Reclamation
Unit 2
Shunters Yard
Station Road
Semley
near Shaftesbury
Dorset SP7 9AH
01747 850350
www.semleyrec.co.uk
Reclamation yard. Stone floors, garden antiques, stone ornaments, fireplaces and surrounds, timber beams and flooring, furniture and doors.

EAST SUSSEX

Ajeer Ltd.
Sugar Loaf Yard
Brightling Road
Woods Corner
Heathfield
East Sussex TN21 9LJ
01424 838555
www.ajeer.co.uk
Reclamation yard.

Best Demolition
Harcourt Lodge Buildings
Burwash Road
Heathfield
East Sussex TN24 8RA
01435 867203
Reclamation yard. Demolition company with a reclamation yard, with flooring, bricks, roof tiles and garden objects.

Danby
Silver How
London Road
Crowborough
East Sussex TN6 1TB
01892 652883
Reclaimed stone. Reclaimed and new stone, architectural elements, Wealden sandstone, landscaping services.

Ecologise Ltd.
Brighton
East Sussex
01273 885458
www.ecologise.co.uk
Ecological general building. Reclaimed wood interiors, flooring, shelves and kitchens, green carpentry and natural decorating. AECB member.

Emmaus Brighton & Hove
Drove Road
Portslade
Brighton
East Sussex BN41 2PA
01273 412093
www.emmaus.org.uk
Second-hand shop. Second-hand furniture, books, electricals, records, clothes, kitchenware, bric-a-brac.

Industrial Air
Unit 4
Beaconsfield Workshops
25 Ditchling Rise
Brighton
East Sussex BN1 4QL
01273 818848
Reclamation yard. Industrial salvage, signage, machines, glass, steel and enamel for collectors, designers and trade.

Original Oak Company Ltd.
Ashlands
Burwash
East Sussex TN19 7HS
01435 882228
www.original-oak.co.uk
Timber and stone. Specialize in supply of reclaimed materials for flooring, especially wood and terracotta, with a back-up joinery service.

ESSEX

Anglia Building Suppliers Ltd.
Waltham Road
Boreham
Chelmsford
Essex CM3 3AY
01245 467505
www.angliabuildingsuppliers.co.uk
Reclamation yard. Sleepers, bricks, paving, architectural antiques, stoneware, surrounds, doors and more.

Architectural Reclaim
New Barn Farm
Cock Lane
High Wood
Essex CM1 3RB
01277 354777
www.architecturalreclaim.com
Reclamation yard. Reclaimed bricks, railway sleepers, floorboards, oak beams, York stone slabs, cast-ironwork, baths, fireplaces and architectural features.

Ashwells Recycled Timber Products Ltd.
Wick Place
Brentwood Road
Bulphan
Upminster
Essex RM14 3TL
01375 892576
www.ashwellrecycling.com
Reclaimed timber and beams. Second-hand timber, hardwood, softwood, tongue-and-groove flooring, beams up to 30 x 30cm, plus railway sleepers.

Blackheath Demolition and Trading
26C Hythe Quay
Colchester
Essex CO2 8JB
01206 794100
blackheathdemo@ntlworld.com
www.blackheathdemolitionandtrading.co.uk
Reclamation yard. Roofing, flooring, timber, doors, windows, garden antiques, sanitary ware.

Courtyard At Debden Antiques
Elder Street
Debden
near Saffron Walden
Essex CB11 3JY
01799 543007
www.debden-antiques.co.uk
Antique gardenware. Garden antiques, statuary, furniture, contemporary, unusual, decorative, sculpture, carving and commissions.

The Fingerplate Company
The Limes
Coles Oak Lane
Dedham
Essex CO7 6DR
08707 650100
www.fingerplates.com
Brass. Victorian and Edwardian carefully made reproduction brass fingerplates.

J.D. Lawrence
Station Road
Maldon
Essex CM9 4LQ
01621 816138
www.jd-lawrence.co.uk
Lead. Reproduction lead garden ornaments, statuary and fountains.

J. Purdy and Sons
rear of Tiptree Heath Garage
Maldon Road
Tiptree
Essex CO5 0QA
01621 893322
Reclamation yard. Slates, bricks, oak beams, stone paving, tiles and other interior pieces.

Victorian Wood Works Contracts Ltd.
Creekmouth
54 River Road
Barking
Essex IGII ODW
020 8534 1000
www.victorianwoodworks.co.uk
Floor layers and restoration. Supply and fitting of reclaimed flooring.

Victorian Wood Works Ltd.
Creekmouth
54 River Road
Barking
Essex IGII ODW
020 8534 1000
www.victorianwoodworks.co.uk
Reclaimed flooring. Old oak, pitch pine, planks, tongue-and-groove wood-strip, woodblock, wide oak floorboards, floors installed, restored or renovated.

GLOUCESTERSHIRE

Antique and Modern Fireplaces
41-43 Great Norwood Street
Cheltenham
Gloucestershire GL50 2BQ
01242 255235
Antique fireplaces. Mainly antique, some repro.

Architectural Heritage Ltd.
Taddington Manor
Taddington
near Cutsdean
Cheltenham
Gloucestershire GL54 5RY
01386 584414
www.architectural-heritage.co.uk
Fine architectural, statuary and garden. Purveyors of architectural antiques and garden statuary.

Cox's Architectural Salvage Yard
10 Fosseway Business Park
Moreton-in-Marsh
Gloucestershire GL56 9NQ
01608 652505
www.coxsarchitectural.co.uk
Reclamation yard. Reclaimed beams, flooring, timber, doors, bathrooms, fire-surrounds, fittings, brassware and decorative items.

The Original Architectural Antiques Co.
Ermin Farm
Cirencester
Gloucestershire GL7 5PN
01285 869222
www.originaluk.com
Architectural antiques. Period and reproduction fireplaces, beams, flooring, troughs, sundials, statuary, gates, urns, doors and quality items of all description.

Minchinhampton Architectural Salvage
Cirencester Road
Aston Down
Stroud
Gloucestershire GL6 8PE
01285 760886
www.catbrain.com
Reclamation yard. Architectural salvage, garden ornaments, fountains, fireplaces, flooring, timber, railings and gates, baths, bricks, quarry tiles, pavers, slates and tiles.

Ronsons Reclamation
Upper Parting
Sandhurst Lane
Sandhurst
Gloucestershire GL2 9NQ
01452 731236
Reclamation yard. Suppliers of architectural effects and reclaimed building materials.

Windmill Architectural Salvage Co. Ltd.
Thumbstone Farm
Tinkley Lane
Nympsfield
Gloucestershire GLIO 3UH
01453 833233
Architectural antiques.

GREATER MANCHESTER

Capital Brick Specialist
Victoria Mills
Highfield Road
Little Hulton
Greater Manchester M38 9ST
0161 799 7555
www.reclaimedbricks.com
Reclaimed bricks. Major stockist of reclaimed hand-made and wirecut bricks, Cheshire pinks, setts, cobbles and other materials.

Emmaus Mossley
Longlands Mill
Queen Street
Mossley
Ashton-under-Lyne
Greater Manchester OL5 9AH
01457 838608
www.emmaus.org.uk
Second-hand shop. Second-hand furniture, books, electricals, records, clothes, kitchenware, bric-a-brac.

InSitu Manchester
Talbot Mill
Ellesmere Street
Hulme
Manchester M15 4JY
0161 839 5525
www.insitumanchester.com
Architectural antiques. Buy, restore and sell fireplaces, doors, cast iron, glass, garden ware, flooring.

InSitu Manchester South
4 Longford Road (off A56)
Stretford
Greater Manchester M32 0HQ
0161 865 2110
www.insituarchitectural.com
Architectural antiques. Fireplaces, stoves, radiators, baths, doors, flooring, lighting, glass, furniture and statuary.

Bruce Kilner
Ashtons Field Farm
Windmill Road
Walkden
Worsley
Greater Manchester M28 3RP
0161 702 8604
Architectural antiques.

Manchester Reclaimers
36 Ellesmere Street
Manchester M15 4JW
0161 839 8316
Architectural antiques. Fireplaces, doors and door furniture, flooring, radiators and sanitary ware.

Pine Supplies
Lower Tongs Farm
Longshaw Ford Road
Smithills
Bolton
Greater Manchester BL1 7PP
01204 841416
www.pine-supplies.co.uk
Reclaimed timber and beams. Beams, flooring, skirting boards, architrave, window sections and doors.

Quality Pine Stripping
Unit 6
Enterprise Park
Reliance Street
Newton Heath
Greater Manchester M40 3AL
0161 682 9708
www.revive-architectural-antiques.co.uk
Architectural antiques. Wood and metal stripping, also architectural antiques and stained glass workshop.

Rarefinder
Flat 2
4 The Beeches
West Didsbury
Greater Manchester M20 2BG
0161 446 1135
Architectural antiques. By appointment only.

HAMPSHIRE

Arc Reclamation Ltd.
1 Upper Downgate Farm
Sandy Lane
Steep Marsh
Petersfield
Hampshire GU32 2BG
01730 231995
Architectural antiques. Reclaimed flooring and timber, windows, doors, fireplaces, pews, kitchens and garden furniture from reclaimed materials.

D. Brant Reclamation Ltd.
Lakeside Garden Centre
Brimpton Common Road
Brimpton Common
Tadley
Hampshire RG7 4RT
01189 813882
Reclamation yard.

Brook Barn Specialists Ltd.
Brook Cottage
Ramsdell
near Basingstoke
Hampshire RG26 5SW
01189 814379
Reclamation yard. Large stock of top quality oak beams, bricks, tiles, slates, paving.

C.G. Comley & Sons Ltd.
Southern Way
Rye Common
Odiham
Hook
Hampshire RG25 1HU
01256 702178
www.comleydemo.co.uk
Dismantling contractors. Demolition and recycling of all salvageable materials, recycled concrete. Also located in Surrey.

Imperial Stone
New Barn Farm
Rake Road
Milland
Hampshire GU30 7JU
01428 741175
www.imperialstone.co.uk
Composition stone. Composition stone urns and statues.

Isle Of Wight Fireplaces
37 Orchard Street
Newport
Isle Of Wight
PO30 1JZ
01983 617136
Antique fireplaces. Fireplaces and architectural antiques. By appointment.

Jardinique
Old Park Farm
Kings Hill
Beech
near Alton
Hampshire GU34 4AW
01420 560055
www.jardinique.co.uk
Antique gardenware. Antique and quality hand-made garden items.

Le Calipel Antiquities
48 Sutherland Road
Southsea
Hampshire PO4 0EZ
02392 815050
Architectural antiques. French and English architectural antiques, including gates, railings, stoves, sanitary ware and furniture.

P.N.C. Reeves Demolition
Hazelwood Farm
Hensting Lane
Owslebury
near Winchester
Hampshire SO21 1LE
01962 777323
Reclamation yard. Reclaimed materials for the preservation of period properties.

Romsey Reclamation
Station Approach
Romsey Railway Station
Romsey
Hampshire SO51 8DU
01794 524174
www.romseyreclamation.com
Reclamation yard. Reclaimed bricks, tiles, slates, railway sleepers, flagstones, telegraph poles, quarry tiles, crash barriers and architectural salvage.

Sanitary Salvage
190 Spring Road
Sholing
Southampton
Hampshire SO19 2QG
02380 433218
www.sanitary-salvage.co.uk
Antique bathrooms. Discontinued and obsolete sanitary ware, 1950s to 1990s.

Stained Glass Studio
14 Exmouth Road
Southsea
Portsmouth
Hampshire PO5 2QL
02392 814728
Stained glass restorer. Original designs and repairs, with a building and carpentry service.

HEREFORDSHIRE

Baileys Home & Garden
The Engine Shed
Station Approach
Ross-on-Wye
Herefordshire HR9 7BW
01989 561931
www.baileyshomeandgarden.com
Architectural antiques. Reclaimed industrial fittings, garden antiques, reproduction period sanitary ware, vintage home accessories and lighting.

Leominster Reclamation and Architectural Salvage
North Road
Leominster
Herefordshire HR6 0AB
01568 616205
info@leorec.co.uk
Architectural antiques. Garden and architectural antiques, radiators, doors, flooring, stone and more.

Whitehall Reclamation
2 Orchard Cottage
Hampton Bishop
Hereford HR1 4LB
01432 870855
Reclamation yard. Reclaimed building materials and architectural antiques; bricks, flooring, baths and radiators.

HERTFORDSHIRE

Barn Fireplaces
Little Heath Farm
Little Heath Lane
Potten End
Berkhamsted
Hertfordshire HP4 2RY
01442 872360
www.barnfireplaces.co.uk
Antique fireplaces. Original Georgian, Victorian and Edwardian fireplaces in stone, wood, marble and iron. Gas effect, etc. Stoves.

Brondesbury Architectural Ltd.
Little Green Street Farm
Green Street
Chorleywood
Hertfordshire WD3 6EA
01923 283400
www.brondesburyarchitectural.com
Architectural antiques. Architectural antiques, doors, baths, fireplaces, radiators, marble restoration, door stripping and stained glass.

Herts Architectural Salvage & Reclamation
1A Shenley Lane
London Colney
St Albans
Hertfordshire AL2 1ND
01727 824111
www.herts-architectural.co.uk
Reclamation yard.
Reclaimed materials and architectural antiques, dismantling contractors.

Thatching and Building Co.
Thatchers Rest
Levens Green
near Ware
Hertfordshire SG11 1HD
01920 438710
Builder. Restoration of listed timber-framed buildings and thatching.

V&V Reclamation
Tree Heritage Nursery
North Road (A119)
Hertford
Hertfordshire SG14 2PW
01992 550941
www.vandv.co.uk
Architectural antiques.
Reclaimed stone, fireplaces, doors, furniture, flooring, planters, statues, bricks, flagstones plus stripping, polishing and blasting.

Wocko The Woodman
Knebworth
Hertfordshire
01438 812237
www.wocko.com
Carpenter and joiner.
Traditional cleft woodwork, mainly from oak and chestnut.

HUMBERSIDE

The Warehouse Antiques
17-20A Wilton Street
Holderness Road
Hull
Humberside HU8 7LG
01482 326559
kevinmarshall@antiquewarehouse.karoo.co.uk
Architectural antiques.
Bathrooms, doors and glass.

ISLE OF MAN

JCK Ltd.
Portview
Balthane Industrial Estate
Ballasalla
Isle of Man
01624 824893
Demolition salvage.
Plant hirers with occasional second-hand timber, steel, stone, slates, topsoil.

KENT

Architectural Stores
55 St John's Road
Tunbridge Wells
Kent TN4 9TP
01892 540368
www.architecturalstores.com
Architectural antiques.
Antique fireplaces, garden statuary, lighting and architectural fittings.

Artisan Oak Buildings Ltd.
Teynham Centre
80 London Road
Teynham
Sittingbourne
Kent ME9 9QH
01795 522212
www.artisanoak.co.uk
Reclamation yard. Old oak specialists, structural and decorative beams, flooring, cutting, shaping and finishing.

Bath Doctor
34 London Road
Faversham
Kent ME13 8RX
01795 591711
www.bath-doctor.co.uk
Bath resurfacing service.
Co-ordinators of the Bath Restoration Council.

Brass Foundry Castings
PO Box 151
Westerham
Kent TN16 1YF
01959 563863
www.brasscastings.co.uk
Brass. Furniture restoration, reproduction antique brass fittings.

Bygones Reclamation Canterbury Ltd.
Nackington Road
Canterbury
Kent CT4 7BA
01227 767453
www.bygones.net
Architectural antiques.
Fireplaces, radiators, rainwater gear, doors, garden furniture, reclaimed bricks, roofing and timber, French wrought-iron gates and stone.

Catchpole & Rye
Saracens Dairy
Jobbs Lane
Pluckley
Kent TN27 0SA
01233 840840
www.crye.co.uk
Antique bathrooms. Original antique sanitary ware, roll-top baths, basins, taps, WCs, fine quality bath enamellers, restorers and reproductions.

Emmaus Dover
Archcliffe Fort
Archcliffe Road
Dover
Kent CT17 9EL
01304 204550
www.emmaus.org.uk

Second-hand shop. Second-hand furniture, books, electricals, records, clothes, kitchenware, bric-a-brac.

Extreme Architecture
Unit 8
The Oast
Hurst Farm
Mountain Street
Chilham
Kent CT4 8DH
01227 738084
www.extremearchitecture.com

Architectural antiques. From England, France and Spain, gates, railings, urns, jardinières, finials, fireplaces, fountains, columns, rotunda, capitals, doors, bronzes, flooring and stairs.

Kamstar Ltd.
Little Dale Workshops
Colliers Green
Cranbrook
Kent TN17 2LS
01580 211428
www.kamstar.net

Reclamation yard. Importers of reclaimed flooring, pavers and tiles.

The Old Radiator Company
Unit 9
Bilting Farm Business Centre
Bilting
near Ashford
Kent TN25 4HA
01233 813355
www.theoldradiator.co.uk

Antique ironwork. Restored cast-iron radiators, valves and accessories, and refurb of clients' own radiators.

Tina Pasco
Waterlock House
Wingham
Canterbury
Kent CT3 1BH
01227 722151
www.tinapasco.co.uk

Antique gardenware. Garden antiques, benches, statuary, fountains, cloches, staddle stones, carved stone remnants, garden tools, decorative items, mirrors, statues, birdbaths, angels.

Rother Reclamation
The Old Tile Centre
Ashford Road
High Halden
near Tenterden
Kent TN26 3BP
01233 850075

Reclamation yard. Bricks, tiles, slates, large quantities of oak beams and architectural salvage.

Symonds Salvage
Colts Yard
Pluckley Road
Dunsfield
Bethersden
near Ashford
Kent TN26 3DD
01233 820724

Reclamation yard. Reclaimed tiles, slates, bricks, oak, doors, windows, stone, garden items, farm tools, furniture and architectural salvage.

Karl Terry Roofing Contractor
The Glyndes
15 The Street
Wittersham
Tenterden
Kent TN30 7EA
01797 270268
www.kentpegs.com

Builder. Lead and chimney work, listed and period properties, slating and mathematical tiling, stone slating and random slating, conservation and renovation specialists.

LANCASHIRE

Acorn Architectural Antiques
near Preston
Lancashire
01995 640591
07817 952550

Architectural antiques. Cast iron, garden, stone, rural bygones, tiles, shop interiors, bars, seats, doors, partitions and panels. Finder service. Deal mainly with trade; by appointment only.

Edward Haes
10 Garden Street
Accrington
Lancashire BB5 1DB
01254 389559
www.haes.co.uk

Architectural antiques. Architectural antiques, salvage, lighting, radiators, sanitary ware, door furniture, fireplaces, flooring, gates, railings and cupboards.

Ribble Reclamation
Ducie Place
off New Hall Lane
Preston
Lancashire PR1 4UJ
01772 794534
www.ribble-reclamation.com

Architectural antiques. Antique architectural items for garden and building design, flags, setts, stone, gates, lampposts, chimneypieces, statuary, troughs, oak beams.

Riverside Reclamation Ltd.
Raikes Clough Industrial Estate
Raikes Lane
Bolton
Lancashire BL3 1RP
01204 533141

Stone. Reclaimed bricks, flagstones, cobbles and other stone flooring, assorted reclaimed timber.

Steptoe's Yard
Park Close Quarries
Moor Lane
Salterforth
Barnoldswick
Lancashire BB18 5SP
01282 813313
www.steptoesyard.co.uk

Reclamation yard. Architectural salvage and reclamation; aka D&D Contractors.

Stonescape (UK) Ltd.
The Stone Centre
Ince Moss Industrial Estate
Cemetery Road
Wigan
Lancashire WN3 4NN
01942 866666

Reclamation yard. Large stocks of reclaimed and new York stone flagstones, granite setts, walling stone, hand-made and wirecut bricks, roof tiles and slates.

Tricklebank Rural Heritage Ltd.
Ormskirk
Lancashire
01704 841831

Developer. Restoration builders using reclaimed materials.

LEICESTERSHIRE

Britain's Heritage
Shaftesbury Hall
3 Holy Bones
Leicester
Leicestershire LE1 4LJ
01162 519592
www.britainsheritage.co.uk

Antique fireplaces. One of the country's best selection of antique fireplaces, some repro. Established 1979.

Corporate Architecture Ltd.
35 Leicester Road
Blaby
Leicestershire LE8 4GR
01162 789257
www.corporatearchitecture.co.uk

Architects.

C.R. Crane & Sons Ltd.
Manor Farm
Main Road
Nether Broughton
Leicestershire LE14 3HB
01664 823366
www.crcrane.co.uk

Builders. Conservation builders and joinery restoration.

Pavilion Estates Ltd.
7 Somerby Road
Pickwell
near Melton Mowbray
Leicestershire LE14 2RA
01664 454869

Developers. Design and building of new buildings using old materials, also restoration of old buildings.

LINCOLNSHIRE

Halton Architectural
Hemswell Antiques Centre
Caenby Corner Estate
Hemswell Cliff
Gainsborough
Lincolnshire DN21 5TJ
01427 668389

Architectural antiques.

R&R Reclamation
Top Farm
Kirton Road
Blyton
Gainsborough
Lincolnshire DN21 3PE
01427 628753
www.rr-reclamation.co.uk

Reclamation yard. Family-run yard, specializing in bricks. Demolition work undertaken.

Windmill Reclamation
Bays 39–43
Beevor Street
Lincoln
Lincolnshire LN6 7DJ
01522 546921
www.windmillreclamation.co.uk

Architectural antiques. Architectural salvage and doors made from reclaimed pine.

LONDON (NORTH)

Antique Oak Flooring Co.
94 High Street
Hornsey
London N8 7NT
020 8347 8222
Reclaimed flooring.
Antique and reclaimed timber flooring.

Architectural Forum
312-314 Essex Road
Islington
London N1 3AX
020 7704 0982
Architectural antiques.
Full range of architectural and garden antiques, reclaimed doors, original radiators and antique chimneypieces.

Retrouvius Reclamation & Design
Trade Warehouse
2A Ravensworth Road
Kensal Green
London NW10 5NR
020 8960 6060
www.retrouvius.com
Architectural antiques.
Warehouse displaying both raw materials and contemporary design items. Selling and sourcing unusual materials for architects and interior designers. Also offer bespoke tables made from reclaimed timbers.

LONDON (SOUTH EAST)

Emmaus Greenwich
226 Elmley Street
Plumstead
London SE18 7NN
020 8316 5398
www.emmaus.org.uk
Second-hand shop. Second-hand furniture, books, electricals, records, clothes, kitchenware, bric-a-brac.

Junction Works
72 Carmichael Road
South Norwood
London SE25 5LX
07973 781798
Architectural antiques.
Radiators, fireplaces, lighting, flooring, pine stripping and restoration.

LASSCO Flooring
41 Maltby Street
Bermondsey
London SE1 3PA
020 7394 2101
www.lassco.co.uk
Flooring. Antique flooring warehouse, contractors, dealers, importers, old oak, Victorian pine, parquet, strip, block, flags, tiles.

LASSCO RBK
Ropewalk
Maltby Street
Bermondsey
London SE1 3PA
020 7394 2102
www.lassco.co.uk
Antique bathrooms. Old radiators, antique bathrooms, kitchens and kitchenalia, allied salvage.

LASSCO Warehouse
41 Maltby Street
Bermondsey
London SE1 3PA
020 7394 2103
www.lassco.co.uk
Architectural antiques.
Doors, woodwork, entranceways, pub fittings, bygones, metalwork, panelling, lighting, architectural salvage.
Trade and overseas buyers welcome.

Wilkinson plc
Head Office & Workshops
5 Catford Hill
London SE6 4NU
020 8314 1080
www.wilkinson-plc.com
Chandelier manufacturers and glass restorers.

LONDON (SOUTH WEST)

Chelminski Gallery
616 King's Road
London SW6 2DU
020 7384 2227
www.chelminski.com
Fine architectural, statuary and garden.
Antique sculpture and garden ornaments.

Zygmunt Chelminski
Studio GE1
2 Michael Road
London SW6 2AD
020 7610 9731
Marble, stone, terracotta.
Restoration and conservation specializing in Coade stone, terracotta, marble, stone, bronze, alabaster.

Christopher Wray
Classic Lighting
591-593 King's Road
London SW6 2YW
020 7751 8701
Repair and restoration of antique lamps.

Drummonds
78 Royal Hospital Road
Chelsea
London SW3 4HN
020 7376 4499
www.drummonds-arch.co.uk
New and reproduction. Antique bathrooms, reclaimed flooring, architectural antiques and reproduction sanitary ware, Spey, Usk and Torridge traditional roll-top baths.

French House
125 Queenstown Road
Battersea
London SW8 3RH
020 7978 2228
Antique furniture. Antique French eighteenth- to nineteenth-century furniture, mirrors, beds, decorative items.

Nicholas Gifford-Mead Antiques
68 Pimlico Road
London SW1W 8LS
020 7730 6233
www.nicholasgiffordmead.co.uk
Antique fireplaces. Fine eighteenth- and nineteenth-century chimneypieces and occasional garden sculpture.

House Hospital
14A Winders Road
London SW11 3HE
020 7223 3179
www.thehousehospital.com
Reclamation yard. Radiators, fireplaces, doors.

Deborah Hurst
4 Gleneldon Mews
London SW16 2AZ
020 8696 0315
Carver and carving restoration. French polishing, carving, gilding, gesso, mirrors and frames, furniture restorer.

Ken Negus Ltd.
90 Garfield Road
London SW19 8SB
020 8543 9266
Mason. Cleaning and restoration of old properties – no individual pieces.

LONDON (EAST)

LASSCO St Michael's
St Michael's Church
Mark Street
off Paul Street
Shoreditch
London EC2A 4ER
020 7749 9944
www.lassco.co.uk
Architectural antiques. London Architectural Salvage and Supply Co. The largest salvage company in London, buying from all over England, selling worldwide.

Reclaimed
Railway Arches
299-301 Montague Road
Leytonstone
London E11 3EX
020 8558 2811
www.reclaimed.uk.com
Timber. All types of architectural and antique timber. Also doors, sanitary ware, radiators, fireplaces and more.

Westland & Company
St Michael's Church
Mark Street
London EC2A 4ER
020 7739 8094
www.westland.co.uk
Architectural antiques. Mostly fireplaces, also antique chimneypieces, fine grates, architectural elements, panelling, paintings and furniture.

LONDON (WEST)

H. Crowther Ltd.
5 High Road
Chiswick
London W4 2ND
020 8994 2326
www.hcrowther.co.uk
Lead. Antique-style lead garden ornaments, statuary, cisterns, fountains, birdbaths, masks, spouts, plaques, urns and planters (statue, lion) and restoration of lead.

McCartney Rose
Tankerton Works
12 Argyle Walk
London WC1H 8HA
020 7833 0404
www.mcrose.co.uk
Civil and structural engineers.

Tower Security
152 New Cavendish St.
London W1 6YL
020 7631 3605
Lock and key restorers. Refurbishment of old locks. Can copy or restore old keys.

Wilkinson plc
Mayfair Showroom
1 Grafton Street
London W1S 4EA
020 7495 2477
www.wilkinson-plc.com
Chandelier manufacturers and glass restorers.

MERSEYSIDE

Heritage Tiling & Restoration Company
PO Box 18
Seaforth Vale
Seaforth
Liverpool
Merseyside L21 0JX
0151 920 7349
www.tiling.co.uk
Floor and wall tilers. Restorers of geometric and encaustic tile floors, mosaics.

MIDDLESEX

Anthemion Ltd.
PO Box 6
Teddington
Middlesex TW11 0AS
020 8943 4000
www.ornamentalantiques.com
Architectural antiques. Antique sourcing, design and restoration services.

The Cast Iron Co. Ltd.
8 Old Lodge Place
Twickenham
Middlesex TW1 1RQ
020 8744 9992
www.castiron.co.uk
New and reproduction ironwork. Architectural metalwork, fencing, benches, street furniture, restoration, reproduction, some antiques.

The Cast Iron Reclamation Co.
23 Waldegrave Road
Teddington
Middlesex TW11 8LA
020 8977 5977
www.perfect-irony.com
Antique ironwork. All types of reclaimed radiators including antique, hospital and panel radiators restored and tested.

Middlesex Glass
Upper Square
South Street
Old Isleworth
Middlesex TW7 7BG
020 8568 7207
www.middlesexglass.co.uk
Antique glass. Suppliers of new and old glass, stained glass, replacement reeded glass, etched glass.

Peco Of Hampton
139 Station Road
Hampton
Middlesex TW12 2AE
020 8979 8310
www.peco-of-hampton.co.uk
Antique doors. Old and new doors and fireplaces, stained glass and marble.

NORFOLK

Aspect Roofing
The Old Mill
East Harling
Norwich
Norfolk NR16 2QW
01953 717777
www.aspectroofing.co.uk
Composition stone. Rare roof tiles and coping stones, repro Redland concrete tiles including Redland 50 and Delta.

Thos. Wm. Gaze & Son
Diss Auction Rooms
Roydon Road
Diss
Norfolk IP22 4LN
01379 650306
www.twgaze.com
Auctioneers. Regular auction sales per year of architectural salvage, garden antiques, statuary, rural and domestic bygones, old advertising; online catalogues usually available on SalvoWEB.

Mongers
15 Market Place
Hingham
Norfolk NR9 4AF
01953 851868
www.mongersofhingham.co.uk
Architectural antiques. Architectural antiques, old bathroom fittings, reclaimed timber, decorative items, old cast-iron radiators and garden items.

Norfolk Reclaim Ltd.
Brancaster Road
Docking
King's Lynn
Norfolk PE31 8NB
01485 518846
www.norfolkreclaim.com
Reclamation yard. Also antiques.

Philpott Demolition
Low Road
Bunwell
Norfolk NR16 1SU
01953 498298
Reclamation yard.

Stiffkey Lamp Shop
Stiffkey
Wells-next-the-Sea
Norfolk NR23 1AJ
01328 830460
enquiries@stiffkeylampshop.co.uk
www.stiffkeylampshop.co.uk

Restored antique lamp fittings and copies of period lamps. Also glass lampshades for both oil and electric lamps, brass and ceramic wall and ceiling switches and a range of low-voltage and candle garden lighting.

NORTH YORKSHIRE

Robert Aagaard & Co.
Frogmire Works
Stockwell Road
Knaresborough
North Yorkshire HG5 0JP
01423 864805

Antique fireplaces. Antique fireplaces, cast inserts and fire baskets, bespoke eighteenth- and nineteenth-century copies.

Heritage Stone Ltd.
Waterfall Farm
Guisborough
North Yorkshire TS14 6PU
01287 635529

Reclaimed stone. Buyers and sellers of stone and other reclaimed items.

Old Flames
30 Long Street
Easingwold
York
North Yorkshire YO6 3HT
01347 821188
www.oldflames.co.uk

Antique fireplaces. Original fireplaces and antique lighting, and other architectural salvage items.

Period Pine Doors
Helderleigh
Easingwold Road
Huby
North Yorkshire YO61 1HJ
01347 811728
enquiries@periodpinedoors.co.uk
www.periodpinedoors.co.uk

Doors. Huge selection of antique doors, from seventeenth-century ledge and brace to modern 1930s, 1940s and 1950s styles.

The Real Wrought Iron Company
Carlton Husthwaite
Thirsk
North Yorkshire YO7 2BJ
01845 501415
www.realwroughtiron.com

Iron and steel. Probably the sole world suppliers of genuine wrought iron. Supply of both puddled and charcoal wrought iron to blacksmiths, for use in restoration of historic ironwork and high quality commissions.

RF Landscape Products
Doncaster Road
Whitley
North Yorkshire DN14 0JW
01977 782240

Reclaimed timber and beams. Railway sleepers, stone paving, setts, cobbles, reclaimed and new.

Rustic Revival Ltd.
The Chapel
Wells Lane
Malton
North Yorkshire YO17 7NX
01653 699977
07900 196635
maggiebearsuk@aol.com
www.rustic-revival.co.uk

Antique pine and curios.

Chris Topp & Co.
Carlton Husthwaite
Thirsk
North Yorkshire YO7 2BJ
01845 501415
www.christopp.co.uk

Ironwork restorers. Supplier of genuine wrought iron through the Real Wrought Iron Co. Ltd. Restorer of antique wrought ironwork and iron artist.

The Yorkshire Range Company
Japonica
Chapel Lane
Halton East
Skipton
North Yorkshire BD23 6EH
01756 710263
info@yorkshireranges.yorks.net
www.yorkshirenet.co.uk/yorkshirerangecompany

Traditional ranges. Makers of cast-iron ranges, restoration, repairs and spares.

White House Antiques
Thirsk Road
Easingwold
York
North Yorkshire YO61 3NF
01347 821479
Reclamation yard.
Doors, woodwork, panelling, bathrooms, radiators, ironwork, fireplaces, garden ornaments and more.

NORTHAMPTONSHIRE

Classic Reclaims Ltd.
Sudborough Road
Brigstock
Kettering
Northamptonshire NN14 3HP
01536 373783
www.classicreclaims.co.uk
Architectural antiques.
Large stock of stone flooring, architectural timber, radiators, all garden and interior achitectural salvage.

N. Oldfield
Piddington Station House
Denton Road
Horton
Northamptonshire NN7 2BG
01604 870788
Thatcher. Reed and straw thatch, free estimates.

Ransfords
Drayton Way
Drayton Fields
Daventry
Northamptonshire NN11 5XW
01327 705310
www.ransfords.com
Reclamation yard.
Reclaimed bricks, slates, tiles, ridge tiles, coping, setts, railway sleepers, stone, quarries, oak beams, chimney pots, doors, floorboards and new flagstones.

Rococo Antiques & Interiors
1 Bridge Street
Lower Weedon
Weedon
Northamptonshire NN7 4PN
01327 341288
www.nevillegriffiths.co.uk
Architectural antiques.
Antique fireplaces, bathrooms, ironwork, reclaimed materials and decorative items by Neville Griffiths.

NORTHUMBERLAND

Borders Architectural Antiques
2 South Road
Wooler
Northumberland NE71 6SN
01668 282475
Architectural antiques.
Architectural and garden antiques, stripping, fireplaces, doors, some timber beams and flooring, repro statuary.

Hutton Stone Company
West Fishwick
Berwick-upon-Tweed
Northumberland TD15 1XQ
01289 386056
Stone. New buff to green grey rubble and dimension stone from Swinton Quarry, some reclaimed stone, some antique stone features and ornament usually in stock.

Jim Railton Auctioneers
Nursery House
Chatton
Northumberland NE66 5PY
01668 215323
www.jimrailton.com
Auctioneers. Auctions usually include garden and architectural lots.

Woodside Reclamation
Woodside
Berwick-upon-Tweed
Northumberland TD15 2SY
01289 331211
www.redbaths.co.uk
Architectural antiques. Full range of architectural antiques and reclaimed building materials including timber and restoration service.

NOTTINGHAMSHIRE

Kilgraney Railway Sleepers
Owthorpe Road
Cotgrave
Nottingham
Nottinghamshire NG12 3PU
07971 914781
www.railwaysleeper.com
Reclaimed timber and beams.

Mark Richens Architectural
Long Acres
Swinderby Road
Collingham
near Newark
Nottinghamshire NG23 7NX
01636 893930
www.markrichensandsons.co.uk
Architectural antiques.
Architectural salvage, reclaimed building materials and antiques for home and garden.

Nottingham Architectural Antiques
St Albans Works
181 Hartley Road
Radford
Nottingham
Nottinghamshire NG7 3DW
01159 790666
www.naar.co.uk
Reclamation yard.
Original period fireplaces, architectural salvage, sanitary

ware, doors, reclaimed building materials, ornamental stone, leaded glass, flooring and timber.

Sherwood Forest Products
Unit 1
Brunel Drive
Newark
Nottinghamshire HG24 2EG
01636 613100
Timber. Importers and manufacturers of hardwood flooring, pre-finished and unfinished. Oak beams, hardwood mouldings, oak doors, stone paving and artefacts.

OXFORDSHIRE

Oxford Architectural Antiques
16-18 London Street
Faringdon
Oxfordshire SN7 7AA
01367 242268
www.oxfordarchitectural.co.uk
Architectural antiques. Fireplaces, doors, sanitary ware, garden furniture and ornaments, windows, radiators, pine, repro ironware.

Pathway Workshop
Dunnock Way
Blackbird Leys
Oxford
Oxfordshire OX4 7EX
01865 714111
www.pathway-workshop.org.uk
Timber. Disability charity making garden products, birdtables, seats, from scrap and reclaimed wood.

Upton Original Wood Co.
Alden Farm
Aldens Lane
Upton
Didcot
Oxfordshire OX11 9HS
01235 851866
www.uptonwood.com
Reclaimed timber and beams. Antique and new flooring, beams and furniture wood.

SHROPSHIRE

Richard Darrah
Brook House
Station Road
Hodnet
Shropshire TF9 3JD
01630 685385
Timber. Riven oak, cleft oak, batten, laths, boards, trenails, tile pegs, pale fences.

North Shropshire Reclamation
Wackley Lodge Farm
Wackley
Burlton
Shrewsbury
Shropshire SY4 5TD
01939 270719
www.old2new.uk.com
Reclamation yard. Reclaimed bricks, beams, stone, pavers, bathroom fittings, fireplaces, garden ornaments, troughs, floorboards.

Priors Reclamation
Unit 65
Ditton Priors Industrial Estate
Ditton Priors
Shropshire WV16 6SS
01746 712450
www.priorsrec.co.uk
Reclaimed flooring. Wood flooring specialists, stripped pine doors and doors to order, new native hardwood floorboards.

Radburnes Landscapes
Duxmoor Farm
Onibury
Craven Arms
Shropshire SY7 9BQ
01584 856342
Landscape designers. Designers and contractors who like using old materials.

SOMERSET

Alice Bear Architecturals
Old Leggs Farm
Podgers Lane
Ilton
Ilminster
Somerset TA19 9HE
01460 259295
Architectural antiques. Reclaimed pine, lead and stone urns, metalwork, fine statuary, trade suppliers of a range of quality repro compo statuary.

Bridgwater Reclamation Ltd.
The Old Co-op Dairy
44A Monmouth Street
Bridgwater
Somerset TA6 5EJ
01278 424636
www.reclaiming.it
Reclamation yard. Dismantling contractors, recycled building materials, Bridgwater Clay roof tile specialists, doors, fireplaces.

Castle Reclamation Ltd.
Parrett Works
Martock
Somerset TA12 6AE
01935 826483
www.castlereclamation.com
Reclamation yard. Architectural antiques, stone fireplaces, oak flooring and furniture, replica flags.

Frome Reclamation & Salvage
Station Approach
Wallbridge
Frome
Somerset BA11 1RE
01373 463919
www.fromerec.co.uk

Reclamation yard. Pine doors, fireplaces, slates, tiles, beams, flooring, antique stoneware, furniture and bygones.

Gardenalia
4 Mile End
London Road
Bath
Somerset BA1 6PT
01225 329949
www.gardenalia.co.uk

Antique gardenware. Garden and architectural antiques.

Glastonbury Reclamation Ltd.
The Old Pottery
Northload Bridge
Glastonbury
Somerset BA6 9LE
01458 831122
mikedash@ukonline.co.uk

Reclamation yard. Building stone and flooring, roofing, chimneys, garden antiques, statuary and sanitary ware.

JAT Environmental Reclamation
Deeside
Belluton
Pensford Hill
Pensford
Somerset BS39 4JF
01761 492906

Reclamation yard. Reclaimed stone, roof tiles, ridge tiles, timber, architectural salvage and reclaimed materials.

Mark Chudley
The Old Station
Great Western Road
Chard
Somerset TA20 1EQ
01460 62800
www.markchudley.com

Shippers. Specialists in packing and shipping architectural antiques worldwide.

Source
11 Claverton Buildings
High Street
Widcombe
Bath
Somerset BA2 4LD
01225 469200
www.sourced.it

Architectural antiques. Decorative and architectural antiques, from old to retro, including 1950s items, English Rose and Boulton Paul kitchens, revolving summerhouses.

South West Reclamation Ltd.
Wireworks Estate
Bristol Road
Bridgwater
Somerset TA6 4AP
01278 444141
www.southwest-rec.co.uk

Reclamation yard. Roof tiles, slates, flooring and architectural salvage.

SOUTH YORKSHIRE

BBR Auctions
Elsecar Heritage Centre
Elsecar
near Barnsley
South Yorkshire S74 8HJ
01226 745156
www.bbrauctions.co.uk

Auctioneers. Regular auctions of advertising, pub and shop fittings, enamel signs, pot lids.

Palmer Timber Products
Unit 11
Roberts Road Business Park
Roberts Road
Balby
Doncaster
South Yorkshire DN4 0RE
01302 366442

Timber. Importers of carved mahogany and teak fireplaces, Gothic features and garden furniture.

Richard Hunter
15 Aughton Avenue
Aughton
Sheffield
South Yorkshire S26 3XB
0114 287 3465
www.figureheads.co.uk

Consultants. Ship's figurehead historian, valuer, international consultant, restorer, writer and archivist (boat/ship figurehead carvings).

South Yorkshire Reclamation
The Yard
Sussex Street
Sheffield
South Yorkshire S4 7YY
0114 272 0874

Reclaimed timber and beams.

Viking Reclamation
Cow House Lane
Armthorpe
Doncaster
South Yorkshire DN3 3EE
01302 835449
www.reclaimed.co.uk

Reclamation yard. Over 200,000 quality reclaimed bricks, period oak flooring, cobble setts, architectural items, garden follies and more.

Williams Architectural Reclaims
The Holly
Blackdyke Lane
Kelfield
Doncaster
South Yorkshire DN9 1AQ
01427 728685
Reclamation yard.

STAFFORDSHIRE

Blackbrook Architectural Antiques
Blackbrook Antiques Village
London Road
Lichfield
Staffordshire WS14 0PS
01543 481450
www.blackbrook.co.uk
Architectural antiques. Antique chimneypieces, garden statuary, fountains, gates, doors, stained glass, bathrooms, lighting, reclaimed bricks, stone and timber.

Cawarden Brick Co.
Cawarden Springs Farm
Blithbury Road
Rugeley
Staffordshire WS15 3HL
01889 574066
www.cawardenreclaim.co.uk
Reclamation yard. All period building materials with some quality reproduction products. Full floor supply and fitting service and floor restoration.

Gardiners Reclaimed Building Materials
Brocksford Street
Fenton
Stoke-on-Trent
Staffordshire ST4 3EZ
01782 334532
www.gardinersreclaims.co.uk
Reclamation yard. Reclaimed roof tiles, bricks, quarry tiles, oak beams, sleepers, cobbles, natural stone, flagging and architectural salvage.

Midlands Slate & Tile
Station Road
Four Ashes
Wolverhampton
Staffordshire WV10 7DG
01902 790473
www.slate-tile-brick.co.uk
Reclamation yard. Mainly trade suppliers of reclaimed building materials, roof tiles, slates, flagstones, floorboards, sleepers.

Les Oakes and Sons
Hales View Farm
Cheadle
Staffordshire ST10 4QR
01538 752126
www.lesoakes.com
Reclamation yard. Architectural reclamation dealers, importers of unique and antique artefacts.

Rayson Reclamation
73 Mcghie Street
Hednesford
Staffordshire WS12 4AH
01543 422202
Reclamation yard. Reclaimed bricks, brick specials, roof tiles and slates, timber and miscellaneous salvage.

Peter Weldon Fine Ceramics
Repton House
126 Princes Road
Hartshill
Stoke-on-Trent
Staffordshire ST4 7JL
01782 876360
www.peterweldon.com
Clay. Repro urns, vases and lamps.

Jim Wise
Hot Lane
Burslem
Stoke-on-Trent
Staffordshire ST6 2BN
01782 714735
Reclamation yard. Reclaimed building materials, bricks, roof tiles and slates, and demolition.

SUFFOLK

3A Roofing Ltd.
The Laurels
Copdock
Suffolk IP8 3JF
01473 730660
07889 680618
www.ipswichroofing.co.uk
Reclaimed roof slates and tiles. Also bricks, reclaimed pegs, pantiles, slates, fittings, vents, chimney pots, bricks, timber, roofing contractors.

Abbots Bridge Reclamation Ltd.
Hall Farm
The Street
Lawshall
Bury St Edmunds
Suffolk IP29 4PA
01284 828081
www.abbotsbridge.com
Reclamation yard. Architectural antiques, oak beams, garden antiques, reclaimed building materials, old and new timber flooring.

Cobar Services
Howes Farm Cottage
Stowmarket Road
Ringshall
Stowmarket
Suffolk IP14 2JA
01473 658435
Timber and stone. Bricks, roofing, floors.

Heritage Reclamation
1A High Street
Sproughton
Ipswich
Suffolk IP8 3AF
01473 748519
www.heritage-reclamations.co.uk

Reclamation yard. Ironmongery, stained glass, lighting, sanitary ware, butler's sinks, garden features, flooring, stoves, fireplaces, doors, radiators, furniture.

Tower Reclaim
Tower Farm
Norwich Road
Mendlesham
Suffolk IP14 5NE
01449 766095
architecturalsalvage.uk.com

Reclamation yard. Reclaimed building materials including oak beams, bricks, floorings, roof tiles, stone paving, garden items and oak-framed barns, aka Preservation in Action.

Treesave Reclamation Ltd.
The Barn
Fysh House Farm
Cuckoo Hill
Bures
Suffolk CO8 5LD
01787 227272
www.buresreclamation.co.uk

Reclamation yard. Period building materials and architectural antiques.

Michael Walton A.R.I.B.A.
Oak Farm Barn
Woodditton Road
Kirtling
Suffolk CB8 9PG
01638 730007

Architects and designers. Reclamation-friendly eco-architect, conservation and VAT specialist.

SURREY

Antique Buildings Ltd.
Dunsfold
near Godalming
Surrey GU8 4NP
01483 200477
www.antiquebuildings.com

Reclamation yard. Immense stocks of ancient oak beams, peg tiles, bricks and dismantled barn frames.

Bioregional Reclaimed
17 Dunster Way
Bedzed
Wallington
Surrey SM6 7DA
020 8404 0647
www.bioregional-reclaimed.com

Reclamation yard. Sourcing and supplying reclaimed construction materials, design advice, structural certification for steel or timber, guaranteed cost savings over equivalent new materials.

Chancellors Church Furnishings
Rivernook Farm
Sunnyside
Walton-on-Thames
Surrey KT12 2ET
01932 252736
www.churchantiques.com

Antique church. Architectural and church antiques, fixtures, furniture and furnishings.

Comley Lumber Centre
70 Wrecclesham Hill
Farnham
Surrey GU10 9JX
01252 716882
www.comleydemo.co.uk/index.htm

Reclamation yard. All types of demolition materials arising from own demo sites, also fencing materials.

Cronin's Reclamation
The Barn
Preston Farm Court
Lower Road
Great Bookham
Surrey KT23 4EF
01372 450450
www.salvoweb.com/dealers/damian-cronin/index.html

Architectural antiques. Reclaimed and new flooring, flagstones, roofing, doors, fireplaces, radiators, brasswork and garden antiques.

Drummonds Architectural Antiques Ltd.
The Kirkpatrick Buildings
25 London Road (A3)
Hindhead
Surrey GU26 6AB
01428 609444
www.drummonds-arch.co.uk

Architectural antiques. Antique garden statuary, decorative items, quality period bathrooms. Bath vitreous enamelling service.

FPS
near Redhill
Surrey RH1 5NB
01737 823325

Builder. Conservation builders usually with long waiting list.

Heritage Reclaimed Brick Co.
24 Willow Lane
Mitcham
Surrey CR4 4NA
020 8687 1896
Reclaimed bricks. Reclaimed bricks, architectural salvage and reclaimed materials.

Patrick Normand
near Guildford
Surrey GU5 0LH
01483 892984
www.housedoctoronline.co.uk
Architectural antiques. Sourcing original features, mainly brasswork and some sanitary ware. By appointment only.

Smiths Architectural Salvage
Unit 9
The Looe
Reigate Road
Epsom
Surrey KT17 3BZ
020 8393 4139
Reclamation yard. Fireplaces, doors, reclaimed flooring bought and sold, architectural items.

Sweerts de Landas
Dunsborough Park
Ripley
Woking
Surrey GU23 6AL
01483 225366
www.sweertsdelandas.com
Fine architectural, statuary and garden. Please phone for an appointment to view the antique garden statuary and ornaments in the eighteenth-century gardens.

Wellers Auctioneers Ltd.
70 Guildford Street
Chertsey
Surrey KT16 9BB
01932 568678
www.wellers-auctions.co.uk
Auctioneer. Some house clearances and architectural salvage.

Woodlands Farm Nursery & Reclamation
The Green
Wood Street Village
near Guildford
Surrey GU3 3DU
01483 235536
Reclamation yard. Reclaimed flagstones, bricks, stone, sinks, pavers, troughs, gates and country garden antiques.

TYNE AND WEAR

Chimneypieces
98 Howard Street
North Shields
Tyne and Wear NE30 1NA
0191 257 2118
www.chimneypieces.co.uk
Antique fireplaces. Antique fireplaces and cast-iron inserts, also chandeliers.

G. O'Brien & Sons Ltd.
Cleadon House
Cleadon Lane
East Boldon
Tyne and Wear NE36 0AJ
0191 537 4332
Demolition. Reclaimed roofing, timber, flooring and other architectural salvage.

Olde Worlde Fireplaces & Architectural Salvage
18 Blandford Square
Newcastle upon Tyne
Tyne and Wear NE1 4HZ
0191 261 9229
Antique fireplaces. Period fireplaces including cast-iron inserts, kitchen ranges, Victorian pine vestibule doors, radiators. Stained glass studio open for commission work and repair.

Shiners of Jesmond
81 Fern Avenue
Jesmond
Newcastle upon Tyne
Tyne and Wear NE2 2RA
0191 281 6474
www.shinersofjesmond.com
Architectural antiques. Large selection of antique fireplaces, architectural antiques, brassware, doors, windows, etc.

Tynemouth Architectural Salvage
Correction House Basement
Tynemouth Road
Tynemouth
Tyne and Wear NE30 4AA
0191 296 6070
www.tynemoutharchitecturalsalvage.com
Architectural antiques. Antique bathrooms, fixtures, fittings, architectural, church items, radiators and door furniture.

WARWICKSHIRE

Phoenix Reclaimed
Kendricks Barn
East House
Church Bank
Binton
Stratford-on-Avon
Warwickshire CV37 9TJ
07836 700452
www.granitesets.co.uk
Stone. Reclaimed granite setts, York stone flagstones and balustrades.

Reids Reclamation
Welford Road
Long Marston
Stratford-on-Avon
Warwickshire CV37 8RA
01789 720027
Reclamation yard. Reclaimed bricks, slates, flags, tiles, oak beams, floorboards, doors, setts.

Source 4 U Ltd.
10 The Holloway (Office)
Market Place
Warwick
Warwickshire CV34 4SJ
01926 498444
www.source4you.co.uk
Reclamation yard. New and reclaimed bricks, stone, slates, tiles, oak beams, flooring, doors, hardware and special items.

Temple Reclamation
61-67 Temple Street
Rugby
Warwickshire CV21 3TB
01788 547353
www.templereclamation.co.uk
Reclamation yard. Reclaimed building materials, unusual reclaimed stone garden items and fountains.

Thomas Crapper & Co. Ltd.
The Stable Yard
Alscot Park
Stratford-on-Avon
Warwickshire CV37 8BL
01789 450522
www.thomas-crapper.com
Antique bathrooms. Makers of Victorian-style sanitary ware, also stocks high quality antique sanitary ware. Established in 1861 and relaunched in 1999.

WEST MIDLANDS

Alscot Bathroom Company
1 Oak Farm
Hampton Lane
Solihull
West Midlands B92 0JB
0121 709 1901
www.alscotbathrooms.co.uk
Antique bathrooms. Suppliers of Victorian, Edwardian and Art Deco sanitary ware, and bathroom restoration services.

Conservation Building Products Ltd.
Forge Works
Cradley Heath
Warley
West Midlands B64 5AL
01384 569551
www.conservationbuildingproducts.co.uk
Reclamation yard. Reclaimed brick, stone, timber, architectural antiques, hard landscaping and garden features.

Coventry Demolition Co.
Ryton Fields Farm
Wolston Lane
Ryton-on-Dunsmore
Coventry
West Midlands CV8 3ES
02476 545051
www.coventry-demolition.co.uk
Reclamation yard. Large (300,000 plus) stocks of bricks, tiles, slates, doors, oak, paving, cobbles, radiators, fireplaces, timber flooring.

Emmaus Coventry & Warwickshire
Unit 2
Gosford Industrial Estate
Gosford Street
Coventry
West Midlands CV3 2DT
01203 228282
www.emmaus.org.uk
Second-hand shop. Second-hand furniture, books, electricals, records, clothes, kitchenware, bric-a-brac.

MDS Ltd.
Unit 14-15
Stechford Trading Estate
Lyndon Road
Stechford
Birmingham
West Midlands B33 8BU
0121 783 9274
www.mdsltd.net
Reclamation yard. Doors made from reclaimed wood, reclaimed flooring, door furniture and architectural antiques.

RBS Oak
Lower Farm
Brandon Lane
Coventry
West Midlands CV3 3GW
024 7663 9338
www.rbsoak.co.uk
Reclaimed timber and beams. Reclaimed timber, flooring and oak, railway sleepers, new and reclaimed cast-iron radiators.

Staffordshire Architectural Salvage
22 Skidmore Road
Coseley
Wolverhampton
West Midlands WV14 8SE
01902 401053
Reclamation yard. Setts, kerbs, flagstones, lampposts, staddles, troughs and gates.

The Original Flooring Company
230A Grange Road
Kings Heath
Birmingham
West Midlands B14 7RS
Reclaimed flooring. Reclaimed floor tiles and encaustic floor tiles.

WEST SUSSEX

Arts & Crafts Home
25A Clifton Terrace
Brighton
West Sussex BN1 3HA
01273 327774
www.achome.co.uk
Various. Originals and replicas from the Arts and Crafts, Gothic Revival and Aesthetic movements.

Bedouin
Partridge Barn
Floodgates Farm
West Grinstead
West Sussex RH13 8LH
01403 711441
www.bedouin.uk.com
Architectural antiques. Decorative antiques, architectural and garden, original and repro furniture.

Cobwebs Salvage
Wyvale Garden Centre
London Road
Handcross
West Sussex RH17 6BA
01444 400406
Architectural antiques. Oak beams, doors, flooring, bricks and tiles, fireplaces and stoves, roof slates and ironmongery.

Country Oak Sussex Ltd.
Little Washbrook Farm
Brighton Road
Hurstpierpoint
West Sussex BN6 9EF
01273 833869
Reclamation yard. Oak beams, oak flooring, antique stone and terracotta flooring, chimneypieces, oak doors, staircases and fittings. Please phone first.

Dorton Demolition
Old Station Goods Yard
Station Road
Burgess Hill
West Sussex RH15 9DG
01444 250330
www.dortondemolition.co.uk
Reclamation yard. Reclaimed flooring, doors, timber, bricks, slates, tiles, radiators.

Heritage Oak Buildings
Benefold Farm House
Petworth
West Sussex GU28 9NX
01798 344066
Antique timber-framed buildings. Oak-framed buildings and acquisition of old properties for conversion.

Norton Garden Structures
The Studio
Chichester Road
Upper Norton
Selsey
West Sussex PO20 9EA
01243 607690
www.nortongardenstructures.co.uk
Timber. Design and build bespoke garden structures including bridges, revolving summerhouses, gazebos, decking, and glass painting.

Peter Robinson
Tangmere Corner
Tangmere
Chichester
West Sussex PO18 0DU
01243 774025
Architectural antiques.

Sotheby's
Summers Place
Billingshurst
West Sussex RH14 9AD
01403 833500
www.sothebys.com
Auctioneers.

Traditional Oak and Timber Co.
Highbrook Sawmills
Hammingden Lane
Highbrook
near Ardingly
Haywards Heath
West Sussex RH16 6SS
01444 892646
Timber. Beams, joinery oak, doors, tables and flooring.

Woodall & Emery
Haywards Heath Road
Balcombe
West Sussex RH17 6PG
01444 811608
www.woodallandemery.co.uk
enquiries@woodallandemery.co.uk
Extensive range of period lighting. Repair and restoration services.

Yapton Metal Co.
Burndell Road
Yapton
near Arundel
West Sussex BN18 0HP
01243 551359
Reclamation yard. Dealers in scrap metals and architectural salvage, cross between Steptoe's and the British Museum. Bronze castings and blacksmith courses often held.

WEST YORKSHIRE

Andy Thornton Architectural Antiques Ltd.
Victoria Mills
Stainland Road
Greetland
Halifax
West Yorkshire HX4 8AD
01422 377314
www.andythorntonltd.co.uk
Architectural antiques. Architectural antiques, décor, church interiors, panelled rooms, repro, garden ornaments, hotel, pub and restaurant refurbishments.

Bingley Antiques
Springfield Farm Estate
Flappit
Haworth
West Yorkshire BD21 5PT
01535 646666
www.bingleyantiques.com
Architectural antiques. Architectural antiques, leaded glass windows, doors, chimney pots, ironwork, stonework, garden furniture and troughs. Photos of stock can be sent same day by email.

Chapel House Fireplaces
Netherfield House
St Georges Road
Scholes
Holmfirth
West Yorkshire HD9 1UH
01484 682275
www.chapelhousefireplaces.co.uk
Antique fireplaces. Antique fireplaces and mantels from 1750 to 1910, workshop restoration service only, not *in situ*.

Old English Timbers
Unit 14 Whitehall Ind Est
Whitehall Rd
Leeds
West Yorkshire LS12 5JB
01132 636002
www.oldenglishtimbers.co.uk
Reclaimed flooring. Flooring from resawn reclaimed timber, also architraves, doors and skirting boards.

W. Machell and Sons Ltd.
Low Mills
Guiseley
Leeds
West Yorkshire
01132 505043
www.machells.co.uk
Architectural antiques. Timber, flooring, stone, fireplaces, doors and windows.

West Yorkshire Architectural Antiques & Salvage Co.
The Griffin
Blackmoorfoot Road
Crossland Moor
Huddersfield
West Yorkshire HD4 5AG
01484 431042
www.architectural-antiques.net
Reclamation yard. Antique fireplaces, radiators, doors, garden antiques, architectural stone, reclaimed timber, slates, flagstones and dismantling.

Wharfedale Roofing and Building
John Mooney
4 Hall Lane
Horsforth
Leeds
West Yorkshire LS18 5JE
01132 584226
07802 397099
Roofing and building work.

WILTSHIRE

Architectural Gates
Mallard
Hoopers Pool
Southwick
Trowbridge
Wiltshire BA14 9NG
01225 766944
07787 771317
www.architectural-gates.com
Iron and steel. Suppliers of new and reclaimed wrought-iron/steel gates, railings, street lamps, pillars. Design service for new architectural ironwork and marble.

Beechfield Reclamation Co.
Hopton Park Trading Estate
Devizes
Wiltshire SN10 2ET
01380 730999
Reclamation yard. Reclaimed building materials and architectural antiques.

Salisbury Demolition Ltd.
35 West Street
Wilton
Wiltshire SP2 0DL
01722 743420
Reclamation yard.

WORCESTERSHIRE

Juro Farm & Garden Antiques
Whitbourne
near Worcester
Worcestershire WR6 5SF
www.juro.co.uk
Antique gardenware. Farm and garden antiques, cider mills, troughs, staddle stones, sundials, carts, statuary and antique furniture.

NJ Radiators
17 Chadwick Bank
Industrial Estate
Stourport
Worcestershire DY13 9QW
01299 250111
www.njradiators.co.uk
Ironwork restorers. Blast cleaning, restoration and sale of cast-iron radiators, pine stripping.

CHANNEL ISLANDS

Longport Properties
PO Box 158
La Plaiderie House
St Peter Port
Guernsey GY1 4EX
01481 728721
Developers. Property developers who reuse reclaimed materials.

Heritage Stonemasonry Ltd.
12 Parcq du Pont Marquet
La Petite Route du Mielle
St Brelade
Jersey JE3 8FB
07797 727921
Mason. Restoration stonemasons with some materials for sale. Specialize in hand-made paints and lime mortars.

NORTHERN IRELAND

Architectural Salvage & Builders
58 Lough Shore Road
Kanarla
Enniskillen
Co Fermanagh BT74 5NH
028 6632 6071
Reclamation yard. Buyers and sellers of all demolition salvage, sandstone, flooring, tiles, old bricks, Bangor blue slates etc.

Rachel Bevan and Tom Woolley
The Old Mill
80 Church Road
Crossgar
Downpatrick
Co Down BT30 9HR
028 4483 0988
www.bevanarchitects.com
Architects and designers. Architects, design advice, sustainability, reuse of old, reclaimed and antique materials. TW also produces the Green Building Digest.

Eco-Products
11 Rathgannon
Warrenpoint
Co Down BT34 3TU
028 417 72402
www.peopleproductsonline.com
Reclamation yard. Reclaimed building materials, including timber, setts, slates and building stone.

John Fyffe Architectural Salvage
4 Jennymount Street
Belfast BT15 3HW
028 9035 1475
Reclamation yard. Architectural antiques and reclaimed building materials, fireplaces, bathrooms.

Danny O'Kane
64 Blakes Road
Castlerock
Coleraine
Co Londonderry BT51 4UE
028 7084 9024
Reclamation yard. Reclaimed building materials including Bangor blue slates.

Wilsons Conservation Building Products
123 Hillsborough Road
Dromore
Co Down BT25 1QW
028 9269 2304
www.wilsons-arcsalve.com
Reclamation yard. Architectural antiques and reclaimed building materials, hardwood flooring, pine beams, quarry tiles, bricks, Bangor blue slates; dismantling contractors.

SCOTLAND

Auldearn Architectural Antiques
Dalmore Manse
Lethen Road
Auldearn
Highland IV12 5HZ
01667 453087
Architectural antiques. Full range of architectural items, plus furniture, linen, china and two joinery and furniture workshops.

Easy
31 West Bowling Green Street
Leith
Edinburgh EH6 5NX
0131 554 7077
www.easy-arch-salv.co.uk
Architectural antiques. Edinburgh architectural salvage yard. Wide range of antique fireplaces, inserts, pews, carved stone, Victorian baths, cast-iron radiators, panelled doors, period lights and more.

Gaia Architects
The Monastery
2 Hart Street Lane
Edinburgh EH1 3RG
0131 557 9191
www.gaiagroup.org
Architects. Reclamation and eco-friendly green architects.

Gibbs Landscape
Millburn Wood
Edinburgh Road
Dolphinton
West Linton
Borders EH46 7AF
01968 682602
Reclamation yard. Reclaimed building materials: paving, cobbles and sandstone; some garden ornaments; agricultural museum.

Glasgow Architectural Salvage
Unit 1
Albion Complex
1394 South Street
Scotstown
Glasgow G14 0DT
0141 958 1113
www.glasgowarchitecturalsalvage.co.uk
Architectural antiques. Architectural antiques including, fireplaces, shutters, baths, pews.

Holyrood Architectural Salvage
146 Duddingston Road West
Edinburgh
Lothian EH16 4AP
0131 661 9305
www.holyroodarchitecturalsalvage.com
Architectural antiques. Antique doors, fireplaces, inserts, bathrooms, radiators, pews, skirting, architrave and architectural features bought and sold.

Tradstocks
Dunaverig
Ruskie
Thornhill
Stirling FK8 3QW
01786 850400
www.tradstocks.co.uk
Reclamation yard. Largest stocks of reclaimed stone in Scotland, flagstones, setts, stone steps, coping, staddle stones. Stone processing.

Woodworks
Pannell Farm
Kilbarchan Road
Bridge of Weir
Renfrewshire PA11 3RN
01505 690062
Restorers of architectural joinery using new or reclaimed timber. Staircases, kitchens, woodturning including spindles.

WALES

Architectural Reclamation
Unit 1, Queensway
Swansea West Industrial Park
Fforesfach
Swansea
Glamorgan SA5 4DH
01792 582222
www.architecturalreclamation.com
Reclamation yard. Specialists in ninteenth- and early twentieth-century architectural antiques, churches and chapels, complete interiors and elements. Reclamation contractors.

ATC (Monmouthshire) Ltd.
Timber House
2 Mayhill Industrial Estate
Monmouth
Gwent NP25 3LX
01600 713036
www.floorsanddecking.com
Reclaimed flooring. Wooden flooring, skirting, panelling, doors, door furniture. Supply, fit, furnish or refurbish.

Cardiff Reclamation
Site 7 Tremorfa Industrial Estate
Rover Way
Cardiff
Glamorgan CF24 2SD
029 2045 8995
Reclamation yard. Architectural antiques and traditional building materials, flagstones, slates, fireplaces, doors, staircases etc.

Celtic Antique Fireplaces
Unit 13
Players Estate
Clydach
Swansea
Glamorgan SA6 5BQ
01792 476047
www.celticfireplaces.co.uk
Antique fireplaces. Antique fireplaces, cast-iron grates, hoods and tile sets.

Dyfed Antiques & Architectural Salvage
The Wesleyan Chapel
Perrots Road
Haverford West
Pembrokeshire
Dyfed SA61 2JD
01437 760496
www.dyfedantiques.com
Architectural antiques. Fire-surrounds, antiques, joinery, doors, windows, leaded glass, roll-top baths, tiles, flagstones, large stocks.

Gallop & Rivers
Ty-r-ash
Brecon Road
Crickhowell
Powys NP8 1SF
01873 811084
www.gallopandrivers.co.uk
Reclamation yard. Stone and timber architectural salvage, flooring, beams, doors and fireplaces. Also sanitary ware, garden antiques, rugs and furniture.

Great Oak Glass
Prospect Farm
Newchapel
Llanidloes
Powys SY18 6JY
01686 411 277
www.greatoakglass.co.uk
Glass. Bill Bleasdale is an artist in stained glass, designing and constructing pieces to commission.

Drew Pritchard Stained Glass and Architectural Antiques
St George's Church
Church Walks
Llandudno
Conwy LL30 2HL
01492 874004
www.drewpritchard.co.uk
Antique glass. Antique stained glass for sale, restorers, conservation, and architectural antiques.

Radnedge Architectural Antiques
Dafen Inn Row
Dafen
Llanelli
Carmarthenshire
Dyfed SA14 8LX
01554 755790
www.radnedge-arch-antiques.co.uk
Architectural antiques. Hundreds of cast-iron fireplaces, huge amounts of timber, flagstones, quarry tiles, gates, railings and a full range of architectural items.

Russell Roberts
44 The High Street
Llanberis
Gwynedd LL55 4EU
07710 603197
Architectural antiques. Mainly slates, decorative items, architectural pieces and terracotta pots.

Elizabeth A. Rose
Glan Aber
123 Chester Road
Mold
Flintshire
Clwyd CH7 1UJ
01352 752310
www.er-design.co.uk
Interior designer. Happy to specify reclaimed materials.

Superseal Architectural Mouldings
3A Siloh Road
Millbrook Industrial Estate
Landore
Swansea
Glamorgan SA1 2NT
01792 701666
www.supersealgrc.co.uk
Composition stone. Concrete work surfaces, porticos, canopies, colonnades, in stock or made to requirements.

Theodore Sons and Daughters (Chimney Pot Man)
Princes Way
off North Road
Bridgend Industrial Estate
Bridgend
Glamorgan CF31 3AQ
01656 648936

Reclamation yard. Flagstones, walling, floorboards, slates, sinks, gutters, chimney pots.

Welsh Salvage Company
Isca Yard
Milman Street
Newport
Gwent NP20 2JL
01633 212945

Architectural antiques. Fireplaces, flooring, timber, doors, stonework, railings and gates, sanitary ware and bathrooms.

GENERAL

Salvo
PO Box 28080
London SE27 0YZ
020 8761 2316
www.salvo.co.uk

The main trading website where you can find UK dealers, their stock and send them all an email with what you're looking for.

www.wantsandoffers.co.uk

A site for private householders to swap and sell architectural and garden items of some value that they no longer need, often to someone locally.

www.salvoMIE.co.uk

A site mainly used by builders to exchange low-value and waste items like topsoil, plasterboard off-cuts and half-tins of paint.

The Society For The Protection Of Ancient Buildings (SPAB)
37 Spital Square
London E1 6DY
020 7377 1644
www.spab.org.uk

Ancient Monuments Society
St Ann's Vestry Hall
2 Church Entry
London EC4V 5HB
020 7236 3934
www.ancientmonumentssociety.org.uk

Period Property UK
www.periodproperty.com

eBay
www.ebay.co.uk

Internet auction website.

Conservation Register
UK Institute for Conservation
702 The Chandlery
50 Westminster Bridge Road
London SE1 7QY
020 7721 8246
register@ukic.org.uk
www.conservationregister.com

Historic Scotland: Scottish Conservation Bureau
Longmore House
Salisbury Place
Edinburgh EH9 1SH
0131 668 8668

The Victorian Society
1 Priory Gardens
Bedford Park
London W4 1TT
020 8994 1019
www.victorian-society.org.uk

The Victorian Society promotes understanding, appreciation and conservation of the architecture and decorative arts of the period.

Twentieth Century Society
70 Cowcross Street
London EC1M 6EJ
020 7250 3857
www.c20society.org.uk

The Twentieth Century Society aims to safeguard the heritage of British architecture and design after 1914.

English Heritage
23 Savile Row
London W1S 2ET
020 7973 3156
www.english-heritage.org.uk

Cathedral Communications Ltd.
High Street
Tisbury
Wiltshire SP3 6HA
01747 871717
www.buildingconservation.com

An organization that provides information on the preservation, conservation and restoration of historic buildings, churches and garden landscapes. Also publishes the annual Building Conservation Directory.

INDEX

AUTHOR'S ACKNOWLEDGEMENTS

The author would like to thank Richard Atkinson at Hodder & Stoughton, Julian Alexander at LAW, Viv Bowler at BBC Books, and Louise Rota and Susan Crook at Wall to Wall. My gratitude and love, as always, also go to Mum, Dad, Ben, Lisa and my husband Alastair.

PHOTOGRAPHIC ACKNOWLEDGEMENTS

The publishers would like to thank the following people for permission to photograph their collections and yards:

Sam Coster of Mongers, Norfolk. Joe and Geraint Gallop of Gallop & Rivers, Crickhowell, Powys. Lawrence Green of InSitu, Manchester. Drew Pritchard of Drew Pritchard Stained Glass and Architectural Antiques, Llandudno. John Rawlinson of The Original Architectural Antiques Co., Cirencester. Anthony Reeve of LASSCO St Michael, London. Steve Tomlin of Minchinhampton Architectural Salvage, Stroud. All photographs ©Andrew Montgomery.

ADDITIONAL PHOTOGRAPHIC SOURCES

©David Clerihew/Living Etc/IPC Media: 6. ©Antony Crolla: 80. ©Dan Duchars/Red Cover: 40–41, 42, 134. ©Andreas von Einsiedel: 147, 176. ©Andrew Lawson/Designer Ivan Hicks: 116, 132. ©Di Lewis/Elizabeth Whiting Associates: 164. ©Neil Lorimer/Elizabeth Whiting Associates: 175. ©Ray Main/Mainstream: 36 (Baileys Home & Garden), 38 (RE Found-Objects), 60, 68 (Baileys Home & Garden), 90 (Designer Vincente Wolfe), 154, 230. ©Marianne Majerus/Isley Walton Manor: 153. ©Paul Massey/Living Etc/IPC Media: 78, 105, 142. ©Mark Scott/Essentials/IPC Media: 66. ©Ken Sparkes: 8, 52. © Wayne Vincent/Red Cover: 188. © Luke White/The Interior Archive: 4, 79. ©Polly Wreford/Marie Claire/IPC Media: 194.

First published in Great Britain in 2005 by Hodder and Stoughton
A division of Hodder Headline

A Hodder & Stoughton Book

By arrangement with the BBC

The Reclaimers accompanies the television series of the
same name produced by Wall to Wall Media Ltd.

10 9 8 7 6 5 4 3 2 1

A CIP catalogue record for this title is available from the British Library

ISBN 0 340 89500 4

Special photography by Andrew Montgomery
For further photographic acknowledgements, see page 237

Designed by Nicky Barneby
Typeset in 12.5/14.5pt Mrs Eaves and Bellamie by Barneby Ltd, London
Colour reproduction by Butler & Tanner
Printed and bound in Great Britain by CPI Bath Press

Hodder Headline's policy is to use papers that are
natural, renewable and recyclable products and made
from wood grown in sustainable forests. The logging and
manufacturing processes are expected to conform to the
environmental regulations of the country of origin

Hodder and Stoughton Ltd
A division of Hodder Headline
338 Euston Road
London NW1 3BH